CW00496342

# T
# SPACE TOY
# Price Guide

*Also by Frank Thompson*

The Corgi Toy Price Guide
The Dinky Toy Price Guide
The Matchbox Toy Price Guide
The Tri-ang, Minic & Spot-On Price Guide

# The
# SPACE TOY
# Price Guide

## Frank Thompson

A & C Black · London

First published 1995 by
A & C Black (Publishers) Limited
35 Bedford Row, London WC1R 4JH

ISBN 0-7136-3998-9

Photos by Bob Wilson of Ashington

A CIP catalogue record for this book is available from the
British Library.

Typeset in 8½ on 9½  Palatino

Printed in Great Britain
by Bell & Bain Limited, Thornliebank, Scotland

# CONTENTS

# DEDICATION

To Gene Roddenberry (Mr Star Trek)

# INTRODUCTION

Space and mankind has of course always been an interesting subject, and exercised the imagination of writers and film makers for many years. Now characters such as the Six Million Dollar Man and Wonder Woman and series like Star Trek and Star Wars have become household names. There have been more fans for Star Trek than for all the other ideas on outer space and science fiction put together!

Space toys have always been collectable and they always will be. Collectors like the many different items, particularly the mysteries and the richness of colours connected with each robot, rocket or module. It was the Americans who first took to collecting space toys in very large numbers, especially after the space speech by President John F. Kennedy in 1961 and getting men to the moon and then to Mars.

Space toys were exciting, and even though tales spread about Far Eastern toys being rubbish (mostly untrue) and some countries like Britain and France refused to allow in such toys or accepted very few quotas, the Japanese succeeded in leaps and bounds.

The Japanese have always been among the greatest toy makers in the world, and as Pierre Boogaerts says in his book *Robots*, 'The Japanese are the great inventors.' The Japanese industry developed rapidly after the Second World War and soon overtook the French and Germans, who had been the supreme masters for decades. My research into the Japanese output of the 1950s, and even more so of the 1960s, was because of the manufacture of the many robots and

rocket-type craft, both clockwork and battery operated. About one to six million models were made between 1950 and 1980, selling at prices of between 50p and £20, although robots kept a steady price of just under £5.

For example, Yonezawa of Tokyo made a robot. Closed up, he presents a rectangular box, but has a clever mechanical design as the body rises from a stem at the base, arms are elevated, the head emerges, eyes blink with a green glow and a three-dimensional TV screen shows a spaceman with direction finding apparatus while the robot moves forward. He is accompanied by a smaller robot. It is a collector's dream – or should I say nightmare? – trying to locate it.

The Americans were also responsible for investing in the Far East and creating new companies, including toy manufacturers. They were like a cottage industry at first and the quality of work was not high, but as the factories grew the quality of toys got better and better. Space toys had become big business.

Most countries started to import the Far Eastern space toys. People went crazy over them. Although I tried to hint that Japanese toys would one day be very collectable indeed, I was ignored and unheeded for a long time!

Like myself, one of the very first people to notice these small robots and other space toys was Jan Steckelenbourg, a Dutch artist and painter who lived in an old house in central France and who was a great collector. Although it is many years since I saw him while playing my accordion in the district, I remember it as if it were yesterday. He was a true inspiration, and possibly he had the very first sizeable collection of space toys. However, I feel I should add that my own collection of toys in the mid-1960s numbered well over 25000, one of the largest in the world. The size of my collection led to my being featured in the local and national press and on radio and TV time after time!

This price guide to space toys is the first of its kind in the world, and the preparation of the text has not been easy. It has been extremely difficult to know sometimes where to place an entry and to make many of the descriptions understandable for collectors. However, I have had great assistance from people like Hugo Marsh of Christie's of London, a toy expert of the highest grade, and several others who do not wish to be named. The manufacturers themselves have sent me letters and catalogues, as well as their good wishes for my task, and such assistance has been invaluable in preparing the entries for Britain, Hasbro, Kenner Parker, MB and Palitoy.

There are no words to describe the personality and charm, not to mention the talents and skills, of Gene Roddenberry, whom I met at the Trekkers Convention at the then Central Hotel, Newcastle upon Tyne in those wonderful early years of the 1980s. He also gave me help and advice, which makes the entries for Star Trek extremely authentic. I would particularly like to thank this man who arrived with his car filled with space comics, books and videos as well as a rare and wonderful collection of space guns and other items.

Finally, I would like to thank everyone who has helped to bring this book together. There are many space toys still out there in the world if one cares to search for them. Some are rare and expensive, but others are still reasonably priced. Whatever your particular interest in space toys and the amount of money in your pocket, I sincerely hope that you will find this book useful and I would welcome comments from users, not just on prices but on any other aspect as well. Happy collecting!

# HINTS FOR COLLECTORS

The best place to find a collector or to start your own collection is a swapmeet or fleamarket. Sometimes they are a combination of the two. Although the major auction houses such as Sotheby's and Christie's are now only too willing to sell diecast toys, most business is still conducted at toy collectors' fairs which have sprung up all over Europe and the United States. Many have become established as annual events, some are monthly and a few are even weekly. In 1976/77 there might have been about 400 swapmeets a year, now there are probably 400 a month in the United Kingdom alone.

National and particularly local newspapers give details of swapmeets and fleamarkets all the time. *The Collector's Gazette, Exchange and Mart*, local radio and television as well, of course, as collectors' shops are all useful sources of information.

Another, and perhaps unexpected, place which attracts collectors is the traction engine rally. At most of these there are trade stands with dealers buying and exchanging models. Many are special promotional models, advertising a particular show. A publication called *The World's Fair* lists all the traction engine rallies to be held.

Similarly, the major agricultural shows attract many model enthusiasts as do meetings held by various car clubs and bus and coach preservation societies. Information on meetings is usually available locally. The possibility of buying a special promotional model at some of these events is an added incentive for attending.

It is as well to remember that a collector should, whenever possible, buy an additional model as these can become exchangeable and thus a better means of finding those items which may be necessary to complete a collection.

# A NOTE ON THE PRICES

Any price guide is certain to cause controversy and especially one in an area of collecting where prices are often rising very rapidly. There will always be variations of opinion, and you may find that a model which is selling for £50 in one place is available for £5 in another. This is not as true as it was, but I hope very much that this guide will set a general standard to be followed.

The prices were as accurate as possible at the time of going to press, but variations may have occurred in the months it has taken for the book to be produced. Often an item is as valuable as the amount the person can afford to pay, and this is particularly true of rare models where cost can become unimportant to the buyer.

# IMPORTANT NOTICE

The prices given in this guide are the prices that you should expect to pay in order to buy an item. They are not necessarily what you should expect to receive when selling to a dealer. Although every care has been taken in compiling this price guide, neither the publishers nor the author can accept any responsibility whatsoever for any financial loss or other inconvenience that may result from its use.

# ABBREVIATIONS

| | |
|---|---|
| MB | Mint Boxed |
| MU | Mint Unboxed |
| GC | Good Condition |
| | |
| MC | Mint Condition |
| GC | Good Condition |
| FC | Fair Condition |
| | |
| DC | Diecast |
| P | Plastic |
| RT | Rubber Tyres |
| TP | Tin Plate |

# BRITAIN

William Britain (1828–1906) was best known to his friends as 'Little Billy' in his early days, but he soon grew up into an upstanding, determined young man who achieved his ambition in building one of the greatest toy firms in the world. He was knighted for his work and, in his old age, was called 'Old Bill' by his friends.

His achievement in British history was the making of toy soldiers for more than 100 years during which the firm's aim was to produce high class goods, so that the name became associated with quality in the minds of the great buying public in every part of the world. The firm manufactured goods that could be bought with the knowledge that into their make-up went that essential 'British characteristic quality' and although its venture into space toys was limited, any collector should be proud to include Britain's models among his or her items. William Britain would have been equally proud of what his company achieved for the world of space.

| MODEL | MB | MU | GC |
|---|---|---|---|
| **F10  Force Cruiser**<br>White with red interior and spacemen in white, red lights and wheels. Price £5.50. Issued 1987. | £50 | £20 | £10 |
| **F11  Force Interceptor**<br>White with two space pilots in white and with red interior cockpit with black separator band. Price £5.50. Issued 1987. | £50 | £30 | £10 |
| **F20  Force Voyager**<br>White with silver scanners, red front roller wheels and red cockpit interior. Red and white stripes on lower rear sides and two spacemen in white suits. Price £5.50. Issued 1987. | £50 | £20 | £10 |
| **F21  Force Commander**<br>White and red with two spacemen in white, one with silver helmet. Price £5.50. Issued 1987. 215mm. | £40 | £20 | £10 |
| **F30  Force Scout Car**<br>White with white and red wheels, red nose, pilot on red seat and silver suit. Price £3.25. Issued 1987. | £30 | £10 | £5 |
| **F40  Strikeforce Boxed Set**<br>White and red with six spacemen. Issued 1987. 380 x 250 x 90mm. Good investment. | £50 | £20 | £10 |

**F41 Force Station**
On silver-blue base with various decals and many components. Issued 1987. 600 x 410mm.

| | | |
|---|---|---|
| £100 | £40 | £20 |

**F60 Winged Raider**
Blue and white. Issued 1987. 215 x 180 x 100mm.

| | | |
|---|---|---|
| £50 | £25 | £15 |

**F70 Orbital Raider**
Blue and white with large wheels and great actions. Issued 1987. 180 x 110 x 85mm.

| | | |
|---|---|---|
| £50 | £20 | £10 |

**R71 Mauler Raider**
Blue and white with orange interior. Pilot in orange suit with blue helmet. Issued 1987. 215mm.

| | | |
|---|---|---|
| £30 | £15 | £7 |

**R80 Dart Raider**
White with blue at front and rear. Spaceman in orange suit with blue helmet. Issued 1987. 105 x 75 x 50mm.

| | | |
|---|---|---|
| £40 | £20 | £10 |

**R90 Raider Force Boxed Set**
White, blue and orange with six spacemen in orange suits and blue helmets. Issued 1987. 380 x 255 x 90mm.

| | | |
|---|---|---|
| £75 | £30 | £15 |

**No. 9110 Space Craft**
Whether blasting into space from its metal tail fin or rolling smoothly over an alien planet, Stargard pilot and crew always remain upright. Yellow and gold with clear plastic hatch, with view of spacemen in gold suits and green helmets and with green ribbed divider on hatch cover. Price £4.40. Issued 1982. 196 x 120 x 83mm.

| | | |
|---|---|---|
| £50 | £20 | £10 |

**No. 9114 Stargard Flight Buggy**
Multifunction flight craft with removable pilot, buggy and computer. Fits together with all Britain space models. Red, yellow and gold with pilot in gold suit. Price £1.40. Issued 1982.

| | | |
|---|---|---|
| £10 | £6 | £2 |

**No. 9115 Space Cannon**
Defend the Stargard galaxy with a continuous barrage from this action packed double-barrel model. A stable platform is provided by the metal tail fins. Safe ammunition included. Price £2.80. Issued 1982. 129 x 117 x 83mm.

| | | |
|---|---|---|
| £20 | £10 | £5 |

**No. 9116 Space Landing Pad**
Propel the Stargard traveller through space on the fine line provided with this unique module, or use it as a crane or winch for transferring vital supplies. Red, lemon and gold. Price £2.80. Issued 1982. 178 x 117 x 83mm.

| | | |
|---|---|---|
| £10 | £5 | £2 |

**No. 9120  Alien Spaceship**
Pilots remain upright in action cruising. Bright green with black pilot suit and red helmet. Has eight positions. Price £5. Issued 1982/83.

| | MB | MU | GC |
|---|---|---|---|
| | £50 | £20 | £10 |

**No. 9125  Alien Space Cannon**
Both barrels fire ammunition. Stabilised on metal fin. Green with spaceman in black suit and red helmet. Orange trim. Price £2.95. Issued 1983.

| | £30 | £15 | £8 |
|---|---|---|---|

**No. 9126  Mutants**
Red and silver or greyish brown and black. Price £3.75. Issued 1983/84.

| | £30 | £20 | £10 |
|---|---|---|---|

**No. 9127  Alien Space Grabs**
Unique spring loaded grabs mounted on metal tail fin. Has the ability to grab and hold. Green and black with orange decor. Price £2.95. Issued 1983.

| | £20 | £10 | £5 |
|---|---|---|---|

**No. 9128  Alien Space Craft**
Green and black with two space drivers in black and red. Price £4.50. Issued 1983.

| | £30 | £15 | £10 |
|---|---|---|---|

**No. 9130  Stargard Cyborgs**
Grey or silver, and black with red weapons. Issued 1983. Price £2.50 for pack of six.

| | £20 | £10 | £5 |
|---|---|---|---|

**No. 9136  Cyborgs Boxed Set**
Gold and green or silver and green and black, although colours can vary. Price £3.50. Issued 1983/84.

| | £30 | £20 | £10 |
|---|---|---|---|

**No. 9140  Stargard Counter Pack**
Contains 48 models. Issued 1983. Good investment.

| | £50 | £20 | £10 |
|---|---|---|---|

**No. 9140A  Stargards**
In the new livery of green and black. Price £2.85. Issued 1983.

| | £10 | £6 | £4 |
|---|---|---|---|

**No. 9146  Stargards and Aliens**
Boxed set of seven models. Issued 1983. 280 x 70 x 50mm.

| | £20 | £10 | £5 |
|---|---|---|---|

**No. 9147  Stargard Boxed Set**
Yellow or gold and red. Price £2.80. Issued 1983/84.

| | £30 | £20 | £10 |
|---|---|---|---|

**No. 9148  Alien Boxed Set**
Green and black with decorative box. Price £3.50. Issued 1983.

| | £30 | £20 | £10 |
|---|---|---|---|

**No. 9150  Alien Counter Pack**
Contains 48 models.

| | £50 | £20 | £10 |
|---|---|---|---|

| | MB | MU | GC |
|---|---|---|---|
| **No. 9159  Stargard Space Accessory Pack**<br>Good investment. | £20 | £10 | £5 |
| **No. 9160  Alien Space Accessory Pack**<br>Selection of space accessories for addition to the Britain space range. Black and green. Price £1. Issued 1983. | £10 | £5 | £2 |
| **No. 9161  Space Cannon Ammunition**<br>Red. Price £1. Issued 1983. | £5 | £3 | £1 |
| **No. 9170  Alien Mutants**<br>Grey and red. Price £2.80. Issued 1983. | £10 | £5 | £2 |
| **No. 9200  Forceguards**<br>Special pack of 42 models. Red and white. Issued 1987. | £30 | £10 | £5 |
| **No. 9247  Cybertron and Rider**<br>Dark blue and red. Issued 1987. 150mm. | £20 | £10 | £5 |
| **No. 9250  Raiders**<br>Counter pack of 48 models with red suits and blue helmets. Issued 1987. | £50 | £20 | £10 |
| **No. 9252  Terror Raiders**<br>Counter pack of 42 models. Orange, green, yellow and blue. Issued 1987. | £40 | £10 | £5 |
| **No. 9297  Muteron and Raider**<br>Green with white tusks and claw feet. Issued 1987. 150mm. | £20 | £10 | £5 |
| **No. 9299  Raider Accessories**<br>Blue and white. Issued 1987. 100 x 100 x 40mm. | £5 | £2 | £1 |
| **No. 9299  Force Accessories**<br>Red, white and grey. Issued 1987. 100 x 100 x 40mm. | £10 | £4 | £2 |

# CORGI

| MODEL | MB | MU | GC |
|---|---|---|---|
| **No. 3  Batmobile and Boat Set** | | | |
| Black and red. From the TV and movie series. Price 19/11. | | | |
| Issued June 1967. Deleted 1982. 267mm. DC/P. | £85 | — | — |
| **No. 4  Bristol Bloodhound Missile and Ramp Set** | | | |
| Authentic livery. Price 9/11. Issued November 1958. | | | |
| Deleted 1961. 226mm. | £75 | — | — |
| **No. 6  Missile and Rocket Set** | | | |
| Thunderbird, Bloodhound, two trolleys, ramp, scanner | | | |
| van and staff car in Bloodhound space livery with decals. | | | |
| Price 12/6. Issued 1959. Deleted 1962. | £200 | — | — |
| **No. 22  James Bond Set** | | | |
| The first issue was for export only, but later the set was | | | |
| released for the home market. Includes a Lotus Esprit | | | |
| with hydroplanes with flip-out fins. Battery of rockets to | | | |
| fire. White with orange marks. The famous Aston Martin | | | |
| specially equipped with retractable machine guns, | | | |
| overrider, rams, rear bullet screen, and of course the | | | |
| famous ejector seat and the unwelcome passenger. Silver | | | |
| with red interior. Finally, the James Bond Shuttle with | | | |
| retractable undercarriage, opening hatches and space | | | |
| satellite with unfolding solar panels. Price £5.95. Issued | | | |
| 1979. Deleted 1982. Length approx. 800mm for all models. | £150 | — | — |
| **No. 23  Spiderman Set** | | | |
| Produced from the TV and comic book series in authentic | | | |
| Spiderman livery. Blue, red, black and white. Includes | | | |
| the Green Goblin with net, Spiderbike with rocket | | | |
| launchers and drop-down bike stand, Spiderbuggy with | | | |
| net, and the Spidercopter and all decals. Price £6.95. | | | |
| Issued 1980. Deleted 1983. | £250 | — | — |
| **No. 40  Batman Gift Set** | | | |
| Authentic Batman logos and liveries. Price £4.50. Issued | | | |
| September 1976. Deleted 1982. | £200 | — | — |
| **No. 107  Batboat** | | | |
| Full Batman livery. Price 95p. Issued 1967. Reissued | | | |
| August 1976. Deleted 1982. 135mm. DC/P. | £40 | £25 | £10 |

**No. 164  Wild Honey Dragster**
Lemon with green tinted windows, black engine, silver
springs and wheels. White plastic chassis, red and white
decals on sides and bonnet. Price 6/11. Issued December
1971. Deleted 1973. 71mm. DC/P.

| | MB | MU | GC |
|---|---|---|---|
| | £50 | £20 | £10 |

**No. 167  US Racing Buggy**
Model with planet intentions for Moon-probe etc. Orange
and silver with gold engine, black exhausts, plastic
bumpers and wheels. Blue lights, spare wheel on roof,
yellow driver with blue helmet. Stars and stripes decals
and the number '7' on doors and bonnet. Price 6/11.
Issued 1972. Deleted 1975. 95mm. DC/P. Ugly
investment.

| | £100 | £60 | £30 |
|---|---|---|---|

**No. 201  The Saint's Volvo**
No. 258 when issued in 1965. Reissued June 1970 (price
5/11). Deleted 1975. 90mm. DC/P.

| | £30 | £20 | £10 |
|---|---|---|---|

**No. 259  Penguin Model**
Unusual model in silver-blue with plastic off-white or
silver engine and lights. Figure in driving seat with black
coat and top hat and yellow and orange umbrella. Price
£1.25. Issued August 1979. Deleted 1982. 95mm. DC/P.

| | £50 | £25 | £15 |
|---|---|---|---|

**No. 261  Spiderbuggy and Green Goblin**
The buggy is in red and blue with Spiderman at the wheel
dressed in red, blue and black. Silver headlights,
sidelights, trim, with red grille, blue wheels and thick
treaded tyres. The Green goblin is caught in the Spider's
Web, held and being dragged along behind the buggy.
With this model you can capture the Green Goblin and
hold him captive. Price £1.65. Issued July 1979. Deleted
1982. 150mm. DC/P.

| | £75 | £40 | £20 |
|---|---|---|---|

**No. 261  James Bond Aston Martin DB5**
Gold body with all the James Bond special features as
seen in a Bond movie. Price 3/6. Issued October 1965.
Deleted 1966. 97mm. DC/P. Two great features: a
retractable cutting blade on each rear wheel, and
revolving front and rear number plates. Other features
include remote control ejector seat, bullet proof screen,
extending overriders and secretly controlled machine
guns. Silver spoked wheels etc. with two figures on front
seat. This exciting car won the first Toy of the Year
Award. Good investment. Also possible to find with the
signature of the actor who played James Bond, Sean
Connery. Original date of issue on box can more than
treble the price.

| | MB | MU | GC |
|---|---|---|---|
| Unsigned | £100 | £60 | £25 |
| Signed | £2000 | — | — |
| With original date of issue on box | £350 | — | — |

**No. 262  Captain Marvel's Porsche**
Space model in metallic pink with yellow and flame red
fire design along top and sides. Also blue, yellow and
white designs and decals on nose, tail and sides. Driver
with yellow helmet and dark blue suit. Price £1.35. Issued
July 1979. Deleted 1980. 120mm.

£45   £25   £10

**No. 263  Captain America Jetmobile**
Smart rocket shaped model in cream with metallic
bronzed wheels, white and blue rings with star on nose
cone. Blue driver. Price £1.35. Issued July 1979. Deleted
1982. 155mm. DC/P.

£45   £30   £15

**No. 264  Incredible Hulk's Truck**
Metallic pink with silver tint, black grille, bumpers and
wheels. Detachable cage and Hulk figure in green with
red shorts. Hulk decals, and designs in pink and white on
yellow and blue background. From the TV series. Price
£1.65. Issued 1979. Deleted 1983. 120mm.

£75   £35   £20

**No. 265  Supermobile Special**
Medium blue with orange-red interior, black and silver
wheels and rods. Rocket firing model. From the
Superman series. Price £2.25. Issued September 1979.
Deleted 1982. 148mm. DC/P. Good investment.

£200   £100   £50

**No. 266  Chitty Chitty Bang Bang**
Authentic livery and decals with the famous character
figures and all working parts. Price 10/6. Issued
November 1968. Deleted 1973. However, most of the
models were completely sold out within a year of release.
102mm. Investment beyond the wildest dreams of most
collectors. Prices may vary.

£750   £200   £100

**Reissue – Chitty Chitty Bang Bang**
After naming this model 'The most fantasmagorical
Corgi toy in the world', it was decided to reissue it in
November/December 1991. A Christmas competition
was organised, open to all Corgi Collector Club
Members. The first prize was a copy of the original Chitty
Chitty Bang Bang Hollywood première colour
programme, signed by many Hollywood greats,
including Edward G. Robinson, Dick Van Dyke and Sally
Anne Howes, and other stars of the movie, such as
Johnny Mathis and Gary Crosby; a signed copy of the
original cast soundtrack album of the film; and of course
the reissued model itself.
Model

£350   £100   £50

Model with signed programme etc.

£5000   —   —

**No. 266  Spider Bike Special**
Black, orange and red with blue and yellow spiderman
decals, rocket launchers complete with drop-down bike
stand. Spiderman figure in red and blue. Price £1.45.
Issued September 1979. Deleted 1984. 115mm. DC/P.

| | £50 | £30 | £15 |

**No. 267  Batmobile**
Black body with red or gold trimlines. Robin and Batman
figures. Another model based on the very successful TV
series. Price 12/11. Issued October 1965. 127mm. DC/P.

| | £60 | £35 | £20 |

**No. 268  Green Hornet and Black Beauty**
Black and green livery and headlights and built-in
immobiliser, flying radar scanner, and driver with Green
Hornet himself in rear seat firing gun. Taken from TV's
most exciting character. Price 8/11. Issued November
1967. Deleted 1972. 127mm. DC/P.

| | £200 | £75 | £35 |

**No. 268  Batbike Special**
Black with red forks and handle bars, thick black wheels
and tyres, and Batman figure. From the TV series which
had to do with outer space as well as Earth. Price £1.45.
Issued October 1978. Deleted 1983. DC/P.

| | £50 | £30 | £20 |

**No. 269  James Bond Lotus Esprit**
White and black. Extended hydroplanes, concealed fins
and remote controlled battery of rockets. From the film
The Spy Who Loved Me. Price £1.95. Issued 1977. Deleted
1982. 120mm. DC/P.

| | £50 | £30 | £20 |

**No. 270  New James Bond Aston Martin DB5**
New, exciting James Bond car with more advanced items
than the above. Metallic silver. Won the Toy Industrial
Award of the year. Price 10/-. Issued 1968. Deleted 1979.
97mm. DC/P.
Unsigned

| | £200 | £100 | £50 |

Signed

| | £3000 | — | — |

**No. 271  James Bond Aston Martin DB5**
Metallic silver with orange-red interior, gold grille and
bumpers. Silver wheels and two figures. Issued May
1978. Deleted 1983. Price £1.95. 130mm.

| | £50 | £25 | £10 |

**No. C271  James Bond Aston Martin**
Silver-blue with black tyres, silver or gold radiator,
mudguards and wheelhubs. Blue or white figure in car.
Price £2.32. Issued 1984. Deleted 1989. 125mm. DC/P.

| | £30 | £25 | £10 |

**No. C278  Dan Dare Supercar**
Although 1981 proved to be a rather quiet year for Corgi,
it was a model such as this that made the advanced

space-type items very popular and best sellers. Metallic red with orange or yellow interior, showing the Dan Dare emblem in purple, red and yellow, with the black eagle on bonnet. Independently operating radiation shield, cockpit hood, retractable wings and rotating radiator grille. Price £2.35. Issued 1981. Deleted 1982. 132mm. DC/P. £150 £75 £40

### No. 283 OSI-DAF City Car
This was truly named 'The Car of the Future'. Red with black roof, off-white interior, plastic wheels, opening doors, boot and special bonnet. Price 6/6. Issued May 1971. Deleted 1971. 73mm. CD/P. £40 £25 £15

### No. 314 Supercat Jaguar XJS-HE
A model many years in advance of its time. Metallic bronze or dark brown with silver plastic wheels, black bumpers, orange or pink interior and opening doors. Price £2.75. Issued 1982. Deleted 1983. 118mm. £60 £40 £20

### No. A320 The Saint's Jaguar
Authentic Saint's livery of white and decals with opening doors and bonnet, as in TV and movie series. Price £1.65. Issued 1978. Deleted 1982. 122mm. DC/P. £50 £30 £20

### No. 336 James Bond Toyota
White with blue interior. Female driver and James Bond figure firing gun and rockets in boot. From the film You Only Live Twice. Price 5/6. Issued November 1967. Deleted 1968. 102mm. £100 £60 £25

### No. 389 Reliant Bond Bug
Another car of the future. Bright orange with off-white interior, silver wheels, headlights, bumpers and trim. Price 4/11. Issued April 1971. Deleted 1978. 67mm. DC/P. £40 £20 £10

### No. 391 James Bond Ford Mustang
This model soon became another best seller. White and red with black roof, silver trim and bumpers. Price 8/11. Issued March 1972. Deleted 1973. 97mm. DC/P. £50 £30 £15

### No. 431 Vantastic Custom Van
This model was for both home and export markets, but the export model has finer quality and is marked 'export'. White with yellow or gold interior. Vantastic decals in pink and gold lettering. Price £1.15. Issued 1977. Deleted 1982. 122mm. £100 £75 £40

**No. 432  Vantastic Van**
Bright yellow with silver plastic grille, hubs and opening
rear doors. Picturesque design with rough rider on motor
cycle and the word 'Vantastic' in white on sides. Price
£1.35. Issued January 1978. Deleted 1978. 122mm. DC/P.   £75   £40   £15

**No. 433  Vanishing Point Van**
Metallic lemon with opening doors at rear, silver hubs
and fine artistic mural on sides, with Vanishing Point
decals. Price £1.35. Issued January 1978. Deleted 1978.
122mm. DC/P. Hard to find.   £350   £150   £100

**No. 435  Superman Van**
Metallic light blue with silver hubs, trim and thick black
tyres. Orange-red interior with opening rear doors and
black steering wheel. Special 'Superman Supervan Pow'
decals for the US export market before the home market
releases. Decals include Superman figure in red, black
and blue on sides of van. From the TV and comic series.
Price equivalent, to home model, £1.35 to £2.35. Issued
1978. Deleted 1982. 122mm. DC/P. Note that with the
series and comic issues of Superman being withdrawn in
1990, 1991 and 1992, this model and all others to do with
Superman will become very highly priced.   £350   £150   £75

**No. 435  Superman Van**
Light silver metallic blue with doors which opened to
reveal Superman's mobile lab, with authentic Superman
figure. Home and export models differ, but if in doubt
consult an expert. From the TV and comic series. Price
£1.65. Issued January 1979 (home market). Deleted 1982.
122mm. DC/P.   £150   £100   £50

**No. 436  Spiderman Van**
Home market. Dark blue with purple, red and blue,
Spiderman figure design and decals on side of van. From
the TV series. Price £1.45. 122mm. DC/P.   £75   £35   £20

**No. 436  Spiderman Van**
Export market. Black and red with Spiderman decals.
Rear opening doors on export model only. Livery richer
and better finished than home model. Has 'For Export
Only' on box. Likely price £1.35 to £2.35. Issued 1978.
Deleted almost at once after limited production. 122mm.
DC/P. Rare.   £250   £100   £50

**No. 497  Man from UNCLE Car**
Space-like features and always involved with rockets and
the planets. Passenger and driver. Price 8/11. Issued
August 1966. Deleted 1972. 108mm. DC/P.
Dark blue and white   £75   £40   £20
All-white. Rare   £500   —   —

**No. C647  Buck Rogers Star Fighter, First Issue**
Metallic silver with swing wings and rocket launchers.
Yellow, white and pink decals. Two metallic blue rockets.
Price £2.25. Issued 1980. Deleted 1983. 150mm. DC/P.

| | MB | MU | GC |
| --- | --- | --- | --- |
| | £50 | £20 | £10 |

**No. C648  NASA Shuttle**
White with black borders and dark interior. Shuttle opens
at top. Price £2.32. Issued 1984. Deleted 1989. 150mm.

| | £50 | £25 | £10 |
| --- | --- | --- | --- |

**No. 649  James Bond Space Shuttle**
Authentic decals from the James Bond series. Retractable
undercarriage, opening hatches and space shuttle
satellite with unfolding solar panels. Price £2.25. Issued
August 1979. Deleted 1980. 156mm.

| | MB | MU | GC |
| --- | --- | --- | --- |
| Off-white or dark silver-blue with black and orange tail, red or cream shuttle. | £60 | £40 | £20 |
| Cream and orange. Rare | £150 | £90 | £40 |

**No. 803  Beatles Yellow Submarine**
This model belongs to the space age category. Yellow and
white complete with authentic Beatles characters. From
the film of the same name. Price 10/6. Issued February
1969. Deleted 1972. 200mm. Should any person be lucky
enough to have the Beatles' signatures on the box, it is
worth a fortune – especially since the tragic death of John
Lennon. Signatures are always worth money if they are
connected with a particular model like this. One of the
most valuable Corgi investments of all time.

| | MB | MU | GC |
| --- | --- | --- | --- |
| Unsigned | £500 | £300 | £100 |
| Signed | £25000 | — | — |

**No. 806  Lunar Bug**
Special cartoon model. Red and white with blue wings.
Issued October 1970. Deleted 1972. Price 5/11. 127mm.

| | £50 | £30 | £15 |
| --- | --- | --- | --- |

**No. 811  James Bond Moon Buggy**
Another space winner. Blue and white with authentic
decals. Price 8/6. Issued May 1972. Deleted 1973. 95mm.
DC/P.

| | £50 | £30 | £15 |
| --- | --- | --- | --- |

**No. 928  Spidercopter**
Dark blue with red spider legs and an eye-shutter
mechanism. Spiderman decals and design. This model
has an amazing flick-out tongue which retracts from the
body after striking. Price £2.25. Issued October 1979.
Deleted 1982. 150mm. DC/P. Limited production.

| | £150 | £75 | £30 |
| --- | --- | --- | --- |

**No. 930  Drax Helicopter**
White or cream with orange lines and interior. James
Bond decals and emblems. Yellow props. Rocket firing
model. Price £1.45. Issued August 1979. Deleted 1982.
156mm.

| | £35 | £20 | £10 |
| --- | --- | --- | --- |

**No. E3019  James Bond Octopussy Set**
Brown and fawn and black with an off-white or silver
plane which appears after flicking a lever on the horsebox
to drop down and launch the tailgate. From the film
Octopussy. Price £2.75. Issued and deleted 1983.

| | MB | MU | GC |
|---|---|---|---|
| | £150 | — | — |

## JUNIORS

**No. J1  NASA Space Shuttle**
White and black with the American flag and the letters
'USA' at rear. Price 40p. Issued 1984. Deleted 1989. 75mm.

| | £20 | £10 | £5 |
|---|---|---|---|

**No. 2  Blake's 7 Liberator**
Grey and black with green dome door. Price 60p. Issued
1979. Deleted 1982. 73mm.

| | £20 | £10 | £5 |
|---|---|---|---|

**No. 5  NASA Space Shuttle**
White with black base with stars and stripes, decals, flags,
retractable hatch and undercarriage. Price 60p. Issued
1980. Deleted 1983. 71mm.

| | £20 | £10 | £5 |
|---|---|---|---|

**No. 6  Daily Planet Copter**
Red and white with red decals on white squares and
white interior. Price 60p. Issued 1979. Deleted 1982.
78mm.

| | £20 | £10 | £5 |
|---|---|---|---|

**No. E13  Buck Rogers Star Fighter**
Silver-blue with blue jet engine. Price 60p. Issued 1980.
Deleted 1983. 76mm.

| | £20 | £10 | £5 |
|---|---|---|---|

**No. E20  Penguinmobile**
Brilliant white with blue engine. Driver with top hat,
white shirt and umbrella. Decal in red, white, blue and
lemon on bonnet in orange circle. Penguin decal at rear.
Price 60p. Issued 1979. Deleted 1982. 73mm.

| | £20 | £10 | £5 |
|---|---|---|---|

**No. E23  Batman Batbike**
Black and lemon with silver engine, forks etc. Rider in
suit and cape as taken from the series. Price 60p. Issued
1979. Deleted 1983. 70mm.

| | £20 | £10 | £5 |
|---|---|---|---|

**No. E32  The Saint's Car**
Brilliant white with whizz wheels. Price 60p. Issued 1978.
Deleted 1982. 76mm.

| | £20 | £10 | £5 |
|---|---|---|---|

**No. E33  Wonder Woman's Car**
Orange and yellow with figure of Wonder Woman in
driving seat. Silver trim, whizz wheels and black seat.
Price 60p. Issued 1979. Deleted 1982. 76mm.

| | £20 | £10 | £5 |
|---|---|---|---|

**No. E40  James Bond Aston Martin**
Metallic silver with opening roof and ejector seat. Red
interior, silver headlights, bumpers and whizz wheels.
Price 60p. Issued 1979. Deleted 1982. 70mm.

|  | £20 | £10 | £5 |
|---|---|---|---|

**No. E41  Space Shuttle**
White, yellow and black. Price 60p. Issued 1979. Deleted
1982. 71mm.

|  | £20 | £10 | £5 |
|---|---|---|---|

**No. E44  Starship Liberator**
Silver-grey, yellow, orange, white and green with whizz
wheels and full silver trim. Price 60p. Issued 1979.
Deleted 1982. 73mm.

|  | £20 | £10 | £5 |
|---|---|---|---|

**No. E47  Superman Van**
Light blue with Superman decals and emblems. From the
TV and comic series. Price 60p. Issued 1978. Deleted 1982.
68mm.

|  | £20 | £10 | £5 |
|---|---|---|---|

**No. E50  Daily Planet Truck**
Metallic red with dark interior and whizz wheels. Planet
decals. Price 60p. Issued 1979. Deleted 1982. 68mm.

|  | £20 | £10 | £5 |
|---|---|---|---|

**No. E56  Spiderman Van**
Dark metallic blue body with Spiderman decals in
orange, red and black. Price 60p. Issued 1979. Deleted
1982. 68mm.

|  | £20 | £10 | £5 |
|---|---|---|---|

**No. E60  James Bond Lotus Esprit**
White with dark windows. Model from the James Bond
series. '007' in large figures. Price 60p. Issued 1977.
Deleted 1982.

|  | £20 | £10 | £5 |
|---|---|---|---|

**No. E69  Batmobile**
Black livery with authentic decals. Price 60p. Issued 1976.
Deleted 1982. 78mm.

|  | £20 | £10 | £5 |
|---|---|---|---|

**No. E73  Bond's Drax Helicopter**
White, black and yellow with authentic decals. Price 60p.
Issued 1979. Deleted 1982.

|  | £20 | £10 | £5 |
|---|---|---|---|

**No. E75  Spider Copter**
Red, blue and black with authentic decals from the series.
Price 60p. Issued 1979. Deleted 1982. 74mm.

|  | £20 | £10 | £5 |
|---|---|---|---|

**No. E78  Batcopter**
Black and pink prop, blades and skis. Authentic Batman
decals in red and yellow. Price 55p. Issued 1978. Deleted
1982. 74mm.

|  | £20 | £10 | £5 |
|---|---|---|---|

| CORGI | MB | MU | GC |
|---|---|---|---|

**No. E80  Marvel Van**
Drab green with black interior. Marvel design with the
figures of Hulk, Spiderman etc. on sides. Price 60p. Issued
1978. Deleted 1982. 68mm.

| | £20 | £10 | £5 |
|---|---|---|---|

**No. E115  James Bond Citroën 2CV**
Bright lemon with pink interior. Also metallic green
(rare). Black grille and whizz wheels. Price 60p. Issued
1981. Deleted 1983. 74mm.

| Bright lemon | £20 | £10 | £5 |
|---|---|---|---|
| Metallic green | £200 | £100 | £50 |

**No. E148  USS Enterprise**
Authentic livery and decals as in the Star Trek series.
Price 60p. Issued 1982. Deleted 1983. 76mm. Good
investment.

| | £100 | £50 | £25 |
|---|---|---|---|

**No. E149  Klingon Warship**
Klingon livery and decals as in the Star Trek series. Price
60p. Issued 1982. Deleted 1983. 76mm. Another best seller
soon snapped up by the very many collectors.

| | £100 | £50 | £25 |
|---|---|---|---|

**No. E200  The Pencil Eater**
Well-advanced model made in limited production, fully
deserving a place among the toys and space of the future.
Various liveries. Issued and deleted 1983. 85mm. Prices
vary, especially as I often told collectors on TV and in the
press about what to look out for in shops. Good
investment, even though it was hard to understand by
the normal toy and model buying public.

| | £100 | £50 | £25 |
|---|---|---|---|

**No. 2506  Supermobile Set**
Blue and silver Supermobile, and Superman van in
metallic silver with Superman decals. Price £1.10. Issued
1980. Deleted 1983.

| | £200 | — | — |
|---|---|---|---|

**No. 2512  NASA Shuttle Space Ship**
This fine shuttle and starship Liberator are in white and
black and blue, silver and lemon. Price £1.10. Issued 1980.
Deleted 1982.

| | £50 | £25 | £10 |
|---|---|---|---|

**No. 2519  Batman Set**
Batmobile and Batboat in authentic livery with decals.
Price 90p. Issued 1980. Deleted 1983.

| | £50 | £25 | £10 |
|---|---|---|---|

**No. 2521  James Bond**
James Bond space shuttle and Drax helicopter in black
and white with decals. Price £1.10. Issued 1980. Deleted
1983.

| | £50 | £25 | £10 |
|---|---|---|---|

| | MB | MU | GC |
|---|---|---|---|

**No. 2529 James Bond Pack**
The James Bond Lotus in white and the Stromberg helicopter in black and golden lemon. Price 90p. Issued 1980. Deleted 1983.

| | £50 | £25 | £10 |

**No. 2538 NASA Shuttle Pack**
The shuttle in black and white with gold interior, plus the star fighter in off-white, blue and orange with red tips. Price £1.10. Issued 1980. Deleted 1983.

| | £50 | £25 | £5 |

**No. 2601 Batman Triple Pack**
Authentic livery and decals. Contains the helicopter, Batboat and figures. Price £2.25. Issued 1977. Deleted 1982.

| | £50 | £25 | £10 |

**No. 3002 Batman Set**
Black, red and blue authentic livery and decals. From the TV and comic series. Price 17/6. Issued 1970. Deleted 1982.

| | £100 | — | — |

**No. 3004 James Bond OHMSS Set**
Includes the Bobsleigh in yellow with black checks and blue driver with orange helmet, and the Bond Volkswagen 1300 in metallic bronze with silver headlights and bumpers. Also includes the Spectre Bobsleigh in black and bronze with brown driver with black helmet. Price 17/6. Issued 1970. Deleted 1976. Fine set.

| | £200 | — | — |

**No. J3019 James Bond Set**
Includes all the authentic planes, models and cars from the film Octopussy. Price £1.16. Issued 1984. Deleted 1989.

| | £20 | £10 | £5 |

**No. 3030 James Bond Again Set**
Contains the Bond Lotus, Jaws telephone van, Stromberg's helicopter, a Mercedes and a speed boat, all in authentic livery and decals. From the film The Spy Who Loved Me. Price £2.25. Issued 1978. Deleted 1980.

| | £200 | — | — |

**No. 3080 Batman Set**
Authentic livery and emblems. At this time the name changed from 'Gift Sets' to 'Play Sets' on the production lines. However, one or two old-style boxes could have been used and if one was, the set would be worth ten times more. Price £3.35. Issued 1980. Deleted 1983.

| | £200 | — | — |

**No. 3081 Superman Set**
Authentic livery and decals. Price £3.35. Issued 1980. Deleted 1983.

| | £200 | — | — |

**No. 3082  James Bond Set**
Authentic livery and decals. Price £3.35. Issued 1980.
Deleted 1983.

| | MB | MU | GC |
|---|---|---|---|
| No. 3082 | £200 | — | — |

## SUPER JUNIORS

**No. 1001  Aston Martin**
Metallic silver with opening roof, ejector seat and
ejectable passenger. Off-white interior, silver grille, lights
and bumpers. Price 7/6. Issued 1970. Deleted 1975.
86mm.

| | MB | MU | GC |
|---|---|---|---|
| No. 1001 | £50 | £25 | £10 |

**No. 1002  Batmobile**
Black and red with decals. Price 8/11. Issued 1970.
Deleted 1982. 85mm.

| | MB | MU | GC |
|---|---|---|---|
| No. 1002 | £50 | £25 | £10 |

**No. 1003  Batboat**
Batman livery and decals. Price 6/6. Issued 1970. Deleted
1982. 83mm.

| | MB | MU | GC |
|---|---|---|---|
| No. 1003 | £50 | £25 | £10 |

**No. 1006  Chitty Chitty Bang Bang**
This super model has to be included in the world of space
and flying models of the future. A rare find and what an
investment, although little did one realise at the time of
its first issue how valuable it would become. Authentic
livery from the film, complete with figures and all decals.
Price 15/-. Issued 1970. Deleted 1976. 86mm.

| | MB | MU | GC |
|---|---|---|---|
| No. 1006 | £250 | £75 | £45 |

# DINKY

| MODEL | MB | MU | GC |
|---|---|---|---|

**No. 100  Lady Penelope's Fab 1**
Pink with firing rocket, silver trim and six wheels. Driver
and Lady Penelope figures inside plastic opening hood.
From the Thunderbird TV series. Price 15/11. Issued
1966. Deleted 1977. 147mm. Prices vary according to
availability, and the model in the cardboard box is worth
treble the model in the plastic bubble pack.

| | £150 | £50 | £30 |
|---|---|---|---|

**No. 100A  Lady Penelope's Fab 1 Special**
Rare white. Lady Penelope figure with driver, pink
interior with silver trim, wheels etc. Price 15/11. Issued
1972, or discovered at this time. Deleted 1977. 147mm.
DC/TP/RT.

| | £750 | £200 | £10 |
|---|---|---|---|

**No. 101  Thunderbird 4**
Dark green with gold. The number '4' in black on sides,
and 'Thunderbird 4' in white on the model. All working
parts. Space car from the Thunderbird TV series. Price
12/11. Issued 1967. Deleted and replaced by third No. 106
in 1974. 143mm. DC/P.

| | £200 | £100 | £50 |
|---|---|---|---|

**No. 102  Joe's Car**
Green and grey with red cab interior, working parts and
rocket. Automatic opening wings and extending tail fins,
flashing engine, exhaust and independent super
suspension. From the Joe 90 TV series. Price 25/11. Issued
1969. Deleted 1976. 139mm. DC/TP/P.

| | £100 | £50 | £25 |
|---|---|---|---|

**No. 102  Joe's Car Special**
Blue, silver and grey. Otherwise as above. Rare.

| | £275 | £100 | £50 |
|---|---|---|---|

**No. 103  Spectrum Patrol Car**
Metallic red with white interior, silver hubs and trim.
From the Captain Scarlet TV series. Price 9/11. Issued
1968. Deleted 1976. 121mm. DC/P.

| | £100 | £50 | £25 |
|---|---|---|---|

**No. 103  Spectrum Patrol Car**
Metallic blue. Otherwise as above. Rare.

| | £395 | £150 | £75 |
|---|---|---|---|

**No. 103  Spectrum Patrol Car**
Silver. Otherwise as first No. 103.

| | £850 | £350 | £150 |
|---|---|---|---|

| | MB | MU | GC |
|---|---|---|---|

**No. 104  Spectrum Pursuit Vessel**
Blue and white with red interior and all rocket/action.
From the Captain Scarlet TV series. Price £1/2/11. Issued
1968. Deleted 1977. 160mm. DC/P.

| | £100 | £50 | £30 |
|---|---|---|---|

**No. 105  Maximum Security Vehicle**
White with red stripes and red interior. From the Captain
Scarlet TV series. Price 13/9. Issued 1968. Deleted 1975.
137mm. DC/P.

| | £100 | £50 | £30 |
|---|---|---|---|

**No. 105  Maximum Security Vehicle**
Grey with black stripes and flashes on sides. Otherwise
as above. Rare.

| | £500 | £200 | £100 |
|---|---|---|---|

**No. 106  The Prisoner Mini-Moke**
Although this model came from a special series, it had
many aspects of coming from another planet. White with
red and white canopy, opening bonnet and spare wheel.
Price 6/9. Issued 1968. Deleted 1971. 73mm. DC/P. Good
investment, especially if you have the autograph of
Patrick McGoohan on the box.

| | £45 | £20 | £10 |
|---|---|---|---|

**No. 106  The Prisoner Mini-Moke**
Cream with blue and white striped canopy. Otherwise as
above. Rare.

| | £250 | £100 | £50 |
|---|---|---|---|

**No. 106  Thunderbird II**
Blue and white with black markings. Also dark blue with
brown and dark yellow markings (rare and worth treble).
Price 15/11. Issued 1974. Deleted 1978. 153mm. DC/P.

| | £100 | £50 | £25 |
|---|---|---|---|

**No. 107  Stripey the Magic Mini**
White, yellow, red and blue with figures of Candy, Andy
and the Bearandas. Price 10/9. Issued 1967. Deleted 1970.
75mm. DC/P.

| | £70 | £30 | £15 |
|---|---|---|---|

**No. 108  Sam's Car**
Beautiful model with space relations which was well
ahead of its time. Gold with a keyless clockwork motor
and automatic drive. Red interior, the first of its kind in
the world of diecast models. Came with a badge,
'Intelligence Network Lapel'. Price 13/9. Issued 1969.
Deleted 1975. 111mm. DC/P.

| | MB | MU | GC |
|---|---|---|---|
| With badge | £80 | £40 | £20 |
| Without badge | £60 | £30 | £10 |

**No. 108  Sam's Car**
Metallic red, silver, green or blue. Otherwise as above.
Rare.

| | MB | MU | GC |
|---|---|---|---|
| Red with badge | £100 | £50 | £30 |
| Red without badge | £70 | £30 | £15 |

| | MB | MU | GC |
|---|---|---|---|
| Silver with badge | £200 | £150 | £100 |
| Silver without badge | £180 | £140 | £90 |
| Green with badge | £400 | £300 | £200 |
| Green without badge | £375 | £275 | £185 |
| Blue with badge | £500 | £400 | £300 |
| Blue without badge | £480 | £380 | £280 |

**No. 309  Star Trek Gift Set**
Contains USS Enterprise and the Klingon Battle Cruiser.
Price £3.75. Issued 1978. Deleted 1980. Good investment.     £750     £400     £100

**No. 351  UFO Interceptor**
Lime green, orange, silver and black with red tipped cap
firing rocket. From Gerry Anderson's UFO television
programme. Price 14/11. Issued 1971. Deleted 1979.
194mm. P.     £125     £50     £25

**No. 354  Shado-2-Mobile**
All-action model in green, orange and silver with the
word 'Shado' and the number '2' in white on sides.
Twelve wheels and heavy tracks, white interior and firing
rocket on roof. Price 17/11. Issued 1971. Deleted 1979.
145mm. DC/P.     £100     £40     £20

**No. 355  Lunar Roving Vehicle**
Blue and red. Thick black plastic wheels with knobbly
treads and two white men of space. Front and rear wheels
steered by pivoting central column. Model astronauts
and simulated solar energy cells. Price 14/11. Issued
1972. Deleted 1975. 114mm. DC/P.     £150     £90     £50

**No. 357  Klingon Battle Cruiser**
Blue and silver. From the Star Trek TV series. Price £1.55.
Issued 1977. Deleted 1979. 220mm.     £150     £90     £50

**No. 358  USS Enterprise**
Silver and orange with black lettering. Fires 'photon
torpedoes' and doors for access to the shuttlecraft. Price
£2.25. Issued 1976. Deleted 1979. 234mm. DC/P. Classic
model, much sought after when the Star Trek series
became an established favourite on television. Good
investment.     £250     £150     £90

**No. 359  Eagle Transporter**
Lime green, red, white and silver with all working parts.
Adjustable feet etc. From the Space 1999 TV series. Price
£2.50. Issued 1975. Deleted 1979. 234mm. DC/P.     £150     £90     £50

| | MB | MU | GC |
|---|---|---|---|
| **No. 360 Eagle Freighter** <br> All-silver body with red trim. Orange and purple parts with adjustable feet. From the Space 1999 TV series. Price £2.50. Issued 1975. Deleted 1979. 222mm. DC/P. | £100 | £50 | £25 |
| **No. 361 Galactic War Chariot** <br> Lime-green or light yellow-green (worth treble). Silver rocket attachments, wheels and six large tyres. Two astronauts in white and orange or bright yellow. Price £2.50. Issued 1978. Deleted 1980. 126mm. | £125 | £90 | £50 |
| **No. 362 Trident Star Fighter** <br> Dark brown or chocolate with dark orange and yellow markings on wings, tail and nose. Rocket which fires by pressing the centre of the fighter. Price £1.47. Issued 1978. Deleted 1980. 170mm. DC/P. | £45 | £30 | £15 |
| **No. 364 NASA Space Shuttlecraft** <br> Model of the famous US NASA shuttlecraft. White and dark blue with US flag markings on sides. Price £4.35. Issued 1978. Deleted 1980. 186mm. DC/P. | £250 | £150 | £75 |
| **No. 367 Space Battle Cruiser** <br> White with blue engine and cockpit interior. Spaceman pilot. Opening clear plastic hatch. Firing rockets with black tips and rocket holders with deep red or orange circles around them. Price £3.25. Issued 1978. Deleted 1980. 187mm. DC/P. | £175 | £100 | £50 |
| **No. 602 Armoured Command Car** <br> Model designed by Gerry Anderson of Thunderbird TV fame. Metallic dark blue or green livery with black wheels. Tracer-projector and super-radar-scanner powered by its own clockwork motor. Price £2.95. Issued 1976. Deleted 1978. 122mm. DC/P. | £125 | £50 | £30 |
| **No. 755 Harpoon for Fab 1 Car** <br> These items came in packets of six as spare harpoons for No. 100. Issued 1966. Deleted 1977. Price 6d per packet. Being realistic can only be found as mint condition. | £10 | — | — |
| **No. 756 Rocket for Fab 1 Car** <br> Made as a spare item for No. 100. Issued 1966. Deleted 1977. Price 6d each. | £5 | £3 | £1 |
| **No. 1027 Lunar Rover Kit** <br> Made from No. 355. Blue and white with front and rear wheels steered by pivoting central control column. Model astronauts and simulated energy cells. Price 75p. Issued 1972. Deleted 1975. | £50 | — | — |

# DENYS FISHER

The firm of Denys Fisher at Thorpe Arch Trading Estate at Wetherby, near Leeds, turned out some of the best high-quality toys in the years leading up to 1978, the year which I believe was their greatest. Sadly, the company like many others is no longer with us, but many of their toys are. The following are among some of the best investments in the world of space.

| MODEL | MB | MU | GC |
|---|---|---|---|
| **Comic Heroes Games Set** | | | |
| The three games, Batman, Superman and Wonder Woman, were sold separately and in sets, in very attractive, strong boxes. Issued 1978. Deleted 1980. Worthwhile investment. | | | |
| Single game | £100 | £40 | £20 |
| Set | £500 | — | — |
| **Comic Heroes 3-D Game** | | | |
| An exciting 3-D game full of comical, colourful illustrations, to try and rescue the Mayor and the Chief of Police from the shadows of Gotham City, but Batman, Robin, Superman and Shazam are confronted by the arch-villains, the Joker, the Riddler, Mr Freeze and, last but not least, the Flying Penguin, who can knock the heroes down. Issued 1978. Deleted 1980. | £200 | £50 | £20 |
| **Purdey Doll** | | | |
| This figure from the Avengers series had a lot to do with defending the Earth against invaders and spies of all kinds. The doll comes with pink leotard, black tights and yellow or gold blouse, and has six costumes, bought separately or as a set. Issued 1978. Deleted 1980. 11½in. Rare and important model to have in your collection. | | | |
| Doll only | £150 | £75 | £40 |
| Single costume | £50 | — | — |
| Set | £500 | — | — |
| **Board Game New Avengers** | | | |
| All the fun and the thrills from the TV series. Issued 1978. Deleted 1980. While a game for the whole family, as a collecting investment the game should be unused. | £50 | £20 | £10 |

### Batmobile Classic

This large and wonderful model in authentic Batman livery and decals was a delight to ride for any child who owned one, even though the investment value went down as soon as it was unpacked. However, anyone lucky enough to have bought one to put away can count their cash. Issued 1978. Deleted 1980.

|  | MB | MU | GC |
|---|---|---|---|
|  | £750 | £250 | £100 |

### Batcycle

Large and beautifully designed pedal-ridable model with authentic Batman livery and markings. It was packed in a corrugated display box, and was suitable for boys and girls from 2 to 5 years of age. Issued 1978. Deleted 1980. Very rare.

|  | MB | MU | GC |
|---|---|---|---|
|  | £500 | £200 | £50 |

### Adventure Sets

Includes three new outfits to take Steve Austin on all kinds of missions in outer space: red suit with white helmet, black rubber boots and gloves, and chest pack respirator; all-white suit with white gloves and boots, and blue check pack; and for undercover missions, casual blue denim suit with brown shoes, pretend wrist radio and portable Bionic support system. Issued 1978. Deleted 1980.

|  | MB | MU | GC |
|---|---|---|---|
|  | £350 | £150 | £50 |

### Bionic Crisis Game

Really connected with Steve Austin. Issued 1978. Deleted 1980. If anyone had the presence of mind to get the star of the series to autograph this game, then keep it in unused and mint condition to ensure a fine investment.

|  | MB | MU | GC |
|---|---|---|---|
| Unsigned | £50 | £20 | £10 |
| Signed | £1000 | — | — |

### Figure No. 2  Bionic Racing Driver

All set to break the world speed record in his high performance supercar. With 3½in. action figure. Issued 1978. Deleted 1980.

|  | MB | MU | GC |
|---|---|---|---|
| Unsigned | £500 | £200 | £50 |
| Signed by Lee Majors | £2000 | — | — |

### Figure No. 3  Bionic Skydiver

Steve Austin is all set to make a death-defying jump from a high altitude. Ready moulded in sky-diver outfit with detachable harness and real working parachute. Authentic Six Million Dollar Man livery. Issued 1978. Deleted almost at once.

|  | MB | MU | GC |
|---|---|---|---|
| Unsigned | £150 | £60 | £30 |
| Signed by Lee Majors | £1500 | — | — |

**Bionic Maskatron**
Metallic man-machine in white and black as well as blue
with a computerised face that he can transform at will
with three masks. He has three interchangeable arms
with gruesome weapon attachments that he can hide
under his shirt. One of the worst enemies of Steve Austin
and his friends. Issued 1978. Deleted 1980. 13in. Rare.  £750  £250  £100

**Bionic Mission Vehicle**
When the Six Million Dollar Man needs to get anywhere
in a hurry, he jumps into the Bionic mission special which
speeds him to any action. The vehicle is in white, red and
blue, with 'Col. Austin, BMV' in black lettering etc. on
sides. Has three wheels for ground travel, while for
taking to the air it features detachable wheel guards and
a shroud at the front end for flying. The command console
in the cockpit has all the communication equipment the
spaceman needs, including a viewing screen and plug-in
cables. The fuel tanks on the side of the vehicle open up
for storage, and by just pressing an activator button can
be quickly ejected. Should conditions not permit landing,
there is a retractable winch and a hook for dangerous
cargo retrieval while the vehicle is hovering overhead.
Issued 1978. Deleted 1980. 20in.  £500  £200  £50

**Bionic Transporter and Repair Station**
One very clever unit in white, black and blue – a great
favourite which quickly sold out. Apart from a
transporter unit for the red uniformed figure of Steve
Austin with all rocket features, open the back and it is
transformed into a revitalisation chamber to recharge the
Bionic system so that he is ready to spring into further
action. Then, when the front is flipped open, the unit
becomes a complete pretend operating theatre for Bionic
surgery. Issued 1978. Deleted 1980. 17½in.  £350  £100  £50

**Jaime Sommers Doll**
This magnificent doll of the Bionic Woman with her
convincingly lifelike long hair and golden dress was a
real gem of a buy at the time, and was awarded the Toy
Fashion Doll Award for 1978. Issued 1978. Deleted or sold
out almost at once.
Unsigned  £50  £20  £10
Signed  £500  —  —

**Bionic Woman Designer Fashions**
Through the tremendous success of action dolls and the
makers of Barbie and Cindy, it was obvious that the
Bionic era was to produce some of the most sought after
playthings of our time. Even though many people were
not aware of the collectable potential, the dolls soon sold
out, and the more they were used the higher the price for
the ones left in their original condition. When Jaime
Sommers is not out fighting the horror monsters of space
she loves to dress up and go out. The dresses were sold
in blister packs. Issued 1978. Deleted almost at once. The
value for anyone who has the complete collection is very
high, although I give individual prices. Nevertheless, the
individual figures represent very good investments.

**Dress No. 1**
Black halter neck dress that ties around the front to show
off her suntan. Her red shoes pick up the red ties of the
shoestring fashion in order to keep her cool as she dances
the night away with Steve Austin.  £50  —  —

**Dress No. 2**
Stunning ankle-length red dress with red shoes.  £50  —  —

**Dress No. 3**
Lovely green evening dress with white pattern and white
shoes.  £50  —  —

**Dress No. 4**
Long summer dress in orange and lemon with black and
white, and white scarf-like stole with lemon sandals.
Long sleeves.  £50  —  —

**Dress No. 5**
Long gold dress with wide sleeves, white frills and white
dots. White shoes.  £50  —  —

**Dress No. 6**
Orange full-length evening dress with white necklace.
Orange shoes.  £50  —  —

**Dress No. 7**
Blue two-piece trouser suit with pretty red-orange and
yellow striped blouse and blue-type sandals. Gold
buttons.  £75  —  —

**Dress No. 8**
Blue smart Cami-top and wide trousers with white shoes.
Ideal for any outdoor event.  £50  —  —

**Dress No. 9**
Lovely dress like something from a Spanish carnival.
Blue, orange and white pattern, wide and long, flowing
out at bottom with matching shoes.                £75       —         —

**Dress No. 10**
Delightful two-piece trouser suit in all white with
matching sandals.                                 £50       —         —

**Dress No. 11**
Orange dark shade two-piece with crossover top and
wide flowing trousers in the gaucho style. Wide belt and
matching shoes. Ideal for riding. Rare.           £150      —         —

**Dress No. 12**
Charming three-tier party dress in pink flowers on white
background with neat white waistcoat and high heel
shoes.                                            £200      —         —

**Dress No. 13**
Very lucky number for the collector who has this
wide-necked, long-sleeved item, wide at the cuffs, with
bandalero trouser-like outfit. Orange or gold with white
dots. Could be ideal as a new-fashion dressing garment.   £250      —         —

**Dress No. 14**
Another wide, long-sleeved, halter neck dress in green or
blue with wide scarf or shawl, again in the Mexican or
Spanish style. Tassels on end of scarf.          £150      —         —

**Dress No. 15**
Another gaucho-style trouser suit, definitely for the
sporting or riding event. Orange or gold with nice
coloured scarf, wide brimmed hat, wide pockets.   £200      —         —

**Dress No. 16**
Black or dark purple long-style afternoon or evening
dress with long scarf and white fur piece. Matching
colour shoes.                                     £150      —         —

**Dress No. 17**
Pink leotard with black tights, high heel shoes and yellow
or gold-type blouse.                              £150      —         —

### Bionic Woman Figure with Mission Purse

This 12in. action figure was dressed in her blue suit and red shoes, and came in a bright pack. What's more she really was Bionic if you roll up the skin on her arm and raise the secret flaps on her thighs to reveal the life-like Bionic modules. She has a Bionic ear which pings when you turn her head from side to side. Red mission purse with wallet, money and decoding chart. Issued 1978. Deleted 1980.

| | MB | MU | GC |
|---|---|---|---|
| Unsigned | £500 | £200 | £50 |
| Signed | £1000 | — | — |

### Bionic Woman Game

Jaime Sommers has a game all of her own, which meant that all her fans could join in and help her through the tough jobs. Issued 1978. Deleted 1980. If any person was thoughtful and lucky enough to get the game signed by her and dated when the game was issued, they are lucky indeed.

| | | | |
|---|---|---|---|
| Unsigned | £50 | £20 | £10 |
| Signed | £500 | — | — |

### Bionic Woman School House

Even in the further and future world one has to have education. Jaime Sommers's very own school house comes complete with desk, files and books, but they are not what they seem. The desk contains a secret panel concealing a special video centre with a buzzer system. The bookcase behind the desk can be turned round to reveal a secret communications centre system, and the files and books are hollowed out to carry messages. There is even a blackboard on which to chalk. The set also includes Jaime's school dress, and requires two batteries. Action figure bought separately. Issued 1978. Deleted 1980. Rare.

| | | | |
|---|---|---|---|
| Set with figure | £750 | — | — |
| Set without figure | £500 | — | — |

### Hair Styling Boutique

Every girl who dreamed of being the Bionic Woman just had to buy this boutique – if any were left. This and the other Bionic Woman collectables brought into the collector's world many women, young and old alike, and, by the end of the 1970s, over 20 million women had gone collection-mad with space items; I have valued many wonderful toys which have been collected and saved by women of all ages. This model had the head and shoulders of the famous Jaime Sommers well in advance of any of the other heads made by toy companies. The original will be hard to find in the Denys Fisher make. It had long-rooted hair in the head for styling, plus soft skin for make-up. The head can be tilted slightly upwards to the light for easier application of make-up. The head and shoulders lift off the base to reveal a tray of cosmetics for eyes and cheeks, lipstick, false eyelashes, make-up applicator, comb and brush, together with curlers, ribbons and rubber bands for styling hair, plus a rare booklet of designs. Issued 1978. Deleted almost at once.

| | MB | MU | GC |
|---|---|---|---|
| Complete set | £1000 | — | — |
| Booklet | £350 | — | — |

### Six Million Dollar Man Game
The Six Million Dollar Man was not only popular with children, but with the huge viewing public in all parts of the world, and many grown-ups became hooked on these games. There are four Bionic assignment tracks on the board, and each player takes the part of the Six Million Dollar Man. Only the real hero gets through. Issued 1978. Deleted 1980.

| | MB | MU | GC |
|---|---|---|---|
| | £50 | — | — |

### Six Million Dollar Man, Critical Assignment Arms and Legs
This is a set of terrific arms and legs with special functions to prepare Steve Austin for extra-tough assignments. The set includes a Neutraliser arm that delivers a karate chop, and a laser arm that shoots out a red beam, while the oxygen supply arm contains a breathing mask with a plug-in Bionic filter. There is a first-aid leg with a hinged panel device, which opens to reveal a repair tool with repairable circuits, and the exploding leg has a trigger to the thigh to activate a Bionic blowout as several leg panels fly off the reveal the internal Bionic mechanisms. Issued 1978. Deleted 1980.

| | MB | MU | GC |
|---|---|---|---|
| | £300 | £100 | £30 |

### Six Million Dollar Man, Figure No. 1  Scuba Diver
Looking for an undersea espionage base, this Steve Austin figure is another addition you should have for the space series. The figure in authentic livery is moulded in a wet suit with detachable flippers. Issued 1978. Deleted 1980. Rare.

| | MB | MU | GC |
|---|---|---|---|
| Unsigned | £500 | — | — |
| Signed by Lee Majors | £2000 | — | — |

### Six Million Dollar Man, New Figure with Bionic Grip and Backpack Radio
The figure is 13in. high with moulded flexible limbs and body, and is dressed in red track suit and shoes. It is even more Bionic than the previous figure, and in addition to his Bionic eye (which he can actually see through), has a Bionic grip action with his right hand, activated by just depressing the Bionic module in his forearm. He will actually grip, then lift the orange, simulated steel beam

---

which is supplied with this set. When the skin is rolled back on his power arm, the removable Bionic module is revealed nestling in his manmade arm. There are other accessories in the form of a helmet backpack radio. This is a crystal radio which actually works without batteries or electricity. Eye-catching box. Issued 1978 in limited number. Deleted almost at once.

| | MB | MU | GC |
|---|---|---|---|
| Empty box | £100 | — | — |
| Complete with box | £500 | — | — |

### Cyberman
The evil silvery monster in a toy size replica. Issued 1978. Deleted 1980.

| | MB | MU | GC |
|---|---|---|---|
| | £200 | £50 | £25 |

### Cybo-Invader and Interceptor
Everyone interested in space models wanted to add this Cybo-Invader to their collection. It is the ultimate weapon that man can build to help Cyborg in his desperate attempt to defend Earth. It has two working rocket launchers and a fused diamond dome that opens to reveal the interceptor and missile. Guns that also fire safe plastic rockets. The interceptor is a seat-like machine made from tough plastic with wings that fold up or out and actually house rockets that fire. The items come together as a set or separately. Issued 1978. Deleted 1980.

| | MB | MU | GC |
|---|---|---|---|
| | £100 | £50 | £20 |

### Cyborg Figure and Cybo-nators
Many men soon became boys when they saw this model in the shops and wanted to own this 9in. man-machine, knowing the investment and scientific potential it could have. Silver and gold with authentic decals. Cyborg came boxed with a Cybo-eliminator liquidator which fired both water and a toy plastic bolt. It had interchangeable limbs, and in seconds you could remove, feet, hands, arms and legs to push on cosmic weapon extensions. Issued 1978. Deleted 1980. Star Wars classic.

| | MB | MU | GC |
|---|---|---|---|
| | £250 | — | — |

### Doctor Who Figure
This figure looked just like the actor Tom Baker with his large floppy hat, jacket, grey trousers and very long scarf. He even carries a sonic screwdriver to get him out of trouble. Issued 1978. Deleted 1980. 9½in. If you were lucky enough to get Tom Baker to sign the box or the scarf, then put your feet up and retire.

| | MB | MU | GC |
|---|---|---|---|
| Unsigned | £350 | £100 | £50 |
| Signed | £2000 | — | — |

### Doctor Who, K9
Doctor Who's canine friend with its computer brain and antennae ears. Just the thing to take with you on any adventure into space. Issued 1978. Deleted 1980.

| | MB | MU | GC |
|---|---|---|---|
| | £250 | — | — |

**Doctor Who, Leela**
Doctor Who's companion with long auburn hair, simulated leather dress and plastic knife. Issued 1978. Deleted 1980. 9½in.    £250    £100    £30

**Doctor Who, the Dalek**
In authentic livery and markings and made of tough plastic, it moves on wheels with rotating head like a Dalek on TV. Issued 1978. Deleted 1980. A treasure to have should anyone have it in mint and boxed condition.    £500    £150    £50

**Doctor Who, the Tardis**
The Doctor's own Tardis that really works, although better known as a police box. Issued 1978. Deleted 1980.    £500    —    —

**Dusty's Playjet**
In white with red fin and middle stripe, with handles to carry the toy. There is the Dusty figure with red apron and yellow and white trolley. Inside the jet you will find a first-class dining compartment with seats round a central table. There is a wardrobe in which to hang all the Dusty figure's super clothes, and the figure can be sold separately. Issued 1978. Deleted 1980.    £100    —    —

**Futuristic Speed-Burner-Cars**
With their futuristic shapes, these ingenious cars came in six different shapes and liveries: blue, green, yellow, orange, purple and red. A race track was available. Issued 1978. Deleted almost at once.
Model    £100    £50    £20
Track    £100    —    —

**The Giant Robot**
Lifelike robot made from tough plastic with gripping claws. Authentic livery and markings. Issued 1978. Deleted 1980.    £300    —    —

**Oscar Goldman and Briefcase**
Fully articulated figure of the action man with his special briefcase. Open it correctly and you will find secret files and a host of special communications equipment, but should the enemy open it incorrectly, a panel explodes to reveal a special simulated burnt out electronics system. Figure has checked black and white jacket, off-white or grey-blue trousers, sky-blue pullover and brown shoes. Issued 1978. Deleted 1980. 13in.    £250    —    —

**Star Wars, Chewbacca Soft Toy**
Denys Fisher made history when the company manufactured the soft toys from the Star Wars series. The first model is in golden brown with a grin on its face. Issued 1978. Deleted 1980.    £250    £100    £30

**Star Wars, C-3PO Kit**
This very collectable galactic model was an instant
winner with everyone. Made up from many parts,
specially moulded and easy to assemble. In authentic
details from the Star Wars series. Issued 1978. Deleted
1980.

£200    £100    £40

**Star Wars, Muton Figure**
Silver and gold as seen in the Star Wars series, this
monster has flexible joints and interchangeable limbs.
The limbs can be instantly removed and replaced by
horrible Muton weapon extensions, the Mutators. There
are two supplied in the set, the Venom Injector and the
Scorch Bore. Issued 1978. Deleted 1980.

£100    £50    £20

**Star Wars, Muton Subforms**
A Muton is a metamorphic being able to transform
himself into three foul forms: silver, pink and black; green
and silver; orange with silver belt. Issued 1978. Deleted
1980. They are a good investment as only a limited
number were made and they are therefore hard to find.

£250    £100    £00

**Star Wars, Darth Vader Pop Bag**
Another idea from this company was to make two large
pop-bang bags, of which the first was the sinister Darth
Vader, wickedly finished in black, white and magenta,
coming from outer space to challenge the galactic heroes.
Made from tough vinyl, packed in a large, colourful box.
Issued 1978. Deleted 1980.

£250    —    —

**Star Wars, R2-D2 Kit**
Issued 1978. Deleted 1980. Snapped up very quickly. Very
rare.

£300    £150    £60

**Star Wars, R2-D2 Pop Bag**
White with all the authentic Star Wars markings and class
colours. For blowing up and punching. Issued 1978.
Deleted 1980. What an investment.

£500    —    —

**Star Wars, R2-D2 Soft Toy**
Blue and white with all markings as in the film series.
Issued 1978. Deleted 1980.

£500    £100    £50

**Stretch Armstrong**
This is the friend you could definitely do with, if and
when you go into space. A winner from his first
appearance. This 13in. man, in corresponding size with
the Stretch Monster, defies imagination. Pull him, stretch
him (regardless of whether it is body, arms or legs), then

watch in sheer amazement as he slowly unravels himself and goes back to normal size. Eye-catching box. Issued 1978. Deleted 1980. Rare.

| | | | |
|---|---|---|---|
| Figure in box | £500 | — | — |
| Empty box | £150 | — | — |

**Stretch Monster**

If ever nightmares were produced on Earth for any reason, it was nothing compared to what would happen in space or beyond. This nasty green monster with his scaly skin, hideous eyes and fearsome fangs was a space creature that many grown-ups and kids alike just had to buy. Grab the limbs of the monster, and pull, pull, pull, twist and turn, even going quite red in the face to do so. He would stretch to an enormous 48 in., and then shrink back again to normal size. Eye-catching box, which is possibly just as valuable as the monster himself. Issued 1978. Deleted 1980.

| | | | |
|---|---|---|---|
| Monster in box | £750 | — | — |
| Empty box | £200 | — | — |

**Turbo Tower of Power**

You can speed Spiderman on his Spiderbike to the scene of devilish deeds simply by pumping the Turbo Tower of Power – no batteries, no track, just fantastic speed. Dark blue or black. Complete with the Batmobile. Issued 1978. Deleted 1980.

| | | | |
|---|---|---|---|
| | £150 | — | — |

# HASBRO

There is no doubt that success in collecting is due to toy makers all over the world, and the firm of Hasbro made history when it placed the Action Man models on the market. The very young are the collectors of tomorrow. They are also the collectors of today, with many parents becoming interested in this investing game purely through the children, regardless of their age. The following models are most certainly worth buying and putting away in a safe place. Just think if you could open your cupboard and look at those Walt Disney toys that were available, regardless of what era you were born in. So collect any of the following and you are unlikely to regret it.

| MODEL | MB | MU | GC |
|---|---|---|---|
| **Aquaspeeders** Assortment of four special defence force vehicles with waterfire, colour change, revealing battle damage on performance cars, in various liveries, although colours can vary: Speedstream, Aquafend, Deluge and Jetstorm. Made 1993. 297 x 210 x 26mm. First-class investment. | £40 | £20 | £10 |
| **Large Turbomaster Autobot** One of the larger Transformer models, a giant leader of the Turbomasters, a high tech truck and trailer command centre. Colourful missile loading system and rapid fire, gravity feed bombing power. Liveries may vary. Issued 1992. 270 x 340 x 100mm. | £50 | £20 | £10 |
| **Medium Turbomaster Autobot** Pink, blue, silver, lemon, red and black. Transforms into a helicopter with spinning blades, 2 firing mechanisms and 6 missiles. Issued 1992. 133 x 250 x 95mm. | £25 | £10 | £5 |
| **Small Turbomasters/Autobots** Assortment of 4 Autobot cars, each with its own firing mechanism and 3 missiles. Variety of colours. Issued 1992/93. 297 x 210 x 26mm. | £20 | £10 | £5 |
| **Rescue Team, More Autobots** Assortment of four Autobot vehicles, including an off-road vehicle with water cannon and an aeroplane with swing wings, tunnel borers and a heavy lift truck, all with handheld weapons. Liveries may vary. Issued 1992/93. | £20 | £10 | £5 |

**Medium Predator Decepticon**
Blue, pink, white, black and green. Main feature is the
independent viewing system located in the large rocket.
When combined with any of the 4 small Predators, it
gives an additional 4 views. It has a spring firing system,
pivoting launch rack and 5 missiles. Issued 1992. £30 £15 £7.50

**Axelerators**
An assortment of vehicles in exotic designs and various
liveries, although colours can vary. Hot colours and
graphics of these four models with dual-purpose engine,
defence modules forming powerful weapons when they
transform to make powerful mighty fighting robots: Hot
Rider, Rapido, Skram and Zap. Issued 1993. 205 x 143 x
25mm. £40 £10 £5

**Air Commandos**
Fully poseable figures with a glider that can fly up to 12
metres, equipped with pivoting machine guns. Issued
1993. 395 x 273mm. The colours alone are worth
collecting. £30 £15 £7

**Battle Helicopters**
Two colourful helicopters that can really fly, one in pink
and the other with blue livery blades and black bodies,
although designs may vary. Pull the ripcord and watch
them take off. Issued 1993. 206 x 60 x 229mm. £30 £15 £5

**Badger Assault Jeep**
Yellow and black with decals. Highly technical vehicle
complete with spring action missile launcher, working
rear suspension, roll cage and more. Issued 1993. 175 x
100 x 230mm. £40 £20 £10

**The Barracuda Torpedo Launching Submarine**
Blue, black and yellow with decals. Has surface dive
action, and comes with torpedoes and positional
stabiliser fins. Issued 1993. 175 x 230 x 75mm. £40 £20 £10

**Battle Commander**
Voice chip includes three different battle commands and
one weapon. Gold, black, bronze, lemon and orange.
Issued 1992. 196 x 270mm. £20 £10 £5

**The Cobra Ice Sabre Snow Vehicle**
White, red and black with rotating three-man cockpit and
spring loaded missile launcher. Issued 1993. 255 x 365 x
100mm. £40 £20 £10

**The Avalanche Arctic Defender**
White and black with large tracks, decals and authentic
'GI Joe' markings. Heavily armed hovercraft with spring
action rapid-fire mine dispenser, three positional missile
ramps and spring action missile gun with driver in
colourful suit. Issued 1993. 325 x 510 x 105mm.    £40    £20    £10

**The Dinosaurs Figure Assortment**
Very colourful, collectable versions of the likely monsters
that one would meet in outer space. Joint venture
between Jim Henson Productions and the Walt Disney
Company. Six figures to collect: Earl Sinclair, Fran
Sinclair, Baby Sinclair, Charlene Sinclair, Robbie Sinclair
and B.P. Richfield. Each figure 4½–5in. high with
articulated waists. Issued 1992. Price is for six figures.    £60    —    —

**Large Predator Decepticon**
The largest Predator which transforms into an awesome
jet bomber. Skyquakes supersight function operates in
both robot and plane form, and when attached to any of
the 4 small Predators it activates an additional 4 views.
Also features a free bomb-drop system and spring loaded
firing mechanism. Multicoloured, although colours can
vary. Issued 1992. 240 x 340 x 100mm. Much sought after
model.    £50    £20    £10

**The Diobots**
Colourful assortment of three Autobots mainly in silver,
with other colours and designs of artistic splendour.
Prehistoric creatures which transform into robots. Issued
in 1991/92. Reissued 1993. 142 x 229 x 95mm. Classic best
seller and a gem of an investment for the collector.    £60    £30    £15

**Small Predators/Deceptions**
Purple, blue, pink, green and black. High quality
assortment of 4 aeroplanes, each with 3 missiles and
firing mechanism. Issued 1992. 297 x 210 x 26mm.    £20    £10    £5

**New Constructions**
Yellow and silver with purple and black decals.
Assortment of six Decepticons. Construction-style
vehicles. Issued 1991/92. 210 x 144 x 42mm.    £30    £15    £10

**The GI Joe Battlewagon**
Black, blue and orange livery with blue wheels and large
thick double tyres. Devastating battery operated model
with motorised shot cannon firing 8 times, with 4-wheel
action and drive. Cannon detaches to become separate
defence outpost with working winch and room for 8
figures. Issued 1993. 335 x 465 x 180mm.    £50    £20    £10

**The GI Joe Battlewagon Special**
Black, red and green with orange striped decor on side
with name, although colours may vary. Otherwise details
exactly as above. £100 — —

**The GI Joe Cobra Hammerhead**
Silver, dark blue, black and orange decor wheels.
Absolutely wonderful model, being six underwater
assault vehicles in one. Decompression chamber and
elevating missile shaft. Super model craft attached in
white and lemon. Issued 1992. 380 x 565 x 130mm. £50 £30 £10

**The GI Joe Earthquake**
Black, red and lemon. Features 3 spring loaded rockets,
launch tubes, earth digging scoop mechanism and 4
missiles. Issued 1993. 325 x 405 x 135mm. £30 £15 £8

**The GI Joe Tracker**
Grey or off-white, dark blue, black, and decor and decals,
with machine guns and rotating gun turrets. Issued 1992.
255 x 130 x 365mm. £40 £20 £10

**GI Joe with Explosive Action**
In the outer Planets with Fantastic figures and vehicles
well prepared for the future in Space. £50 £20 £10

**GI Joe Figure Assortment 1**
GI Joe and Cobra recruits, now armed with great, exciting
spring loaded weapons. Includes 10 figures. Issued 1993.
245 x 152mm. £20 £10 £5

**GI Joe Figure Assortment 2**
Another 10 colourful figure. Issued 1993. 245 x 152mm. £40 £10 £5

**GI Joe Figure Assortment 3**
A further 10 colourful figures. Issued 1993. 245 x 152mm. £20 £10 £5

**GI Joe Supersonic Fighters**
Contains 6 figures in colourful livery with 4 electronic
sounds and flashing lights. Issued 1993. 196 x 270mm. £20 £10 £5

**The Ninja Figure Assortment**
Outfitted in classical-style Ninja attire with specialised
spring action martial arts. Issued 1993. 245 x 152mm. £30 £15 £5

**James Bond Jr Sports Car**
Flame-red with large black wheels secured with large silver wing-fasteners, black bucket seats, and motifs and decor. James Bond's nephew keeps one step ahead of SCUM with this fired-up street machine which features a side ejector seat and pop-out front black guns with saw blades. Two spring loaded rear exhaust pipe missile launchers and pull-out hub-cap shredder blades. Issued 1993. 230 x 330 x 102mm. — £50 £20 £10

**James Bond Jr Figure Assortment 1**
In a colourful array of liveries, the first six figures include JBJ Scuba Gear, JBJ Street Gear, Dr No, Walker D. Plank, IQ and Odd Job. Fully poseable and come with a unique action mechanism. Issued 1993. Each figure 4½in. high. — £20 — —

**James Bond Jr Figure Assortment 2**
Again in colourful livery, six more figures include JBJ Ninja, JBJ Parachute, Gordo Leita, Jaws, Dr Derange and Mr Buddy Mitchel. Issued 1993. 235 x 191mm. — £20 — —

**The Subcycle**
Yellow, white, silver and black. Sleek vehicle to keep James Bond Jr hot on the trail of his foes on sea or land. Quickly converts from a high tech motorcycle into a well-armed turbo sub. Issued 1993. 203 x 280 x 102mm. — £30 £20 £10

**Monster Face**
Create your own monster if you purchase two of these items, one to keep in mint boxed condition for your collection and one to experiment with. Make and remake this frightful face with over 30 ghoulish parts. Child activated levers articulate jaws and shifty eyes while you think of the outer planets. Issued 1993. — £50 — —

**Obliterators**
One bright multicoloured Decepticon and one bright Autobot, featuring multifiring missile launchers. Colours may vary. Issued 1993. Hasbro Special. — £40 £20 £10

**Obliterators Clench**
Two colourful vehicles, a fire-fighting Simba 8 x 8 Hasbro Special and a Decepticon 18-wheel Trucker/Clench. Issued 1993. 240 x 290 x 100mm. — £50 £20 £10

**The SCUM Shark**
Deep purple, orange and black. Fast, lightweight and powerful with ripcord action. Room for one figure. Spring loaded canopy, missile and removable turbo engine. Issued 1993. Super investment. — £40 £20 £10

**The ATF Storm Eagle**
Fawn with blue interior cockpit. GI Joe air jet with
realistic water blaster-nose, cannon, retractable landing
gear and flip- down inflight handle. Issued 1993. 413 x
255 x 95mm.    £20    £10    £5

**The ATF Liquidator Jet Fighter**
Black with blue interior. Realistic water blasting nose and
cannon. 413 x 255 x 95mm. Issued 1993.    £20    £10    £5

**Sinclair the Talking Baby Monster**
Another model which surely belongs to the world of the
future, even though it could have appeared from the past
after being found under a large volcanic rock. Believe it
or not, this is a huggable version of Baby Sinclair from the
TV series. Soft body and vinyl head. It says six phrases
with the pull of a string. 225 x 300 x 380mm. Issued
1991/92. Reissued 1993.    £75    —    —

**Skyscorchers**
Assortment of sky-machines in an array of bright liveries
and exotic designs, hot colour graphics, cunningly
disguised undercarriage radar-like targeting systems
which detach to form separate weapons: Tornado,
Terradive, Snipe and Hawk. Issued 1993. 205 x 143 x
25mm.    £40    £15    £10

**Stormtroopers**
In various colour-livery range, four deadly vehicles
ready for any attack or defence with waterfire and colour
change revealing battle damage on concept-type cars:
Drench, Rage, Hydradread and Aquablast. Issued 1993.
297 x 210 x 26mm.    £40    £20    £10

**Transformers, a Breed of the Future**
Collectable Autobots from another planet.    £40    £20    £10

**Trakkons-Decepticons**
These Autobots, Ironfist and Deftwing, and the
Trakkons, Fearswoop and Calcar come in wonderfully
varied colours. Issued 1993. 120 x 250 x 95mm.    £40    £20    £10

**Trakkons-Lightformers**
Orange, black, green, yellow and red, although colours
can vary with these 2 Autobot and 2 Decepticon land
vehicles and jets. Super accurate tracer with rotating gun
barrel and noise. Issued 1993.    £40    £20    £10

# KENNER PARKER

In 1993 the advert balloons of Kenner Parker gave indication of some wonderful top quality playthings for people of all ages. Although meant for children, many grown-up collectors were given the chance to invest in Batman, Superman and the Terminator, and get involved in the planets. I would like to thank the company management for their sincere help in my research, and I hope that you will find that the following provide a real chance for investment.

| MODEL | MB | MU | GC |
|---|---|---|---|
| **The Batwing Special** <br> Metallic blue or black (rare and worth treble). Large disc silver wheels and motifs on model. The model has a claw to capture villains and the option of the extra fire power supplied by spring loaded missile launcher on the wings. Issued June 1993. | £20 | £10 | £5 |
| **Animation Batmobile** <br> All-new look for the Badmobile. Dark blue livery and motifs with blue tinted clear cockpit. Place Batman in the cockpit and use the pull-out wheel shredders to tear up trouble if Batman needs an aerial advantage. Issued June 1993. 206 x 94 x 432mm. | £30 | £10 | £5 |
| **Batman the Animated Series** <br> Colourful 3-D game which features all of Batman's foes including the Joker, Penguin and Catwoman. Colourful box. Issued 1993. | £20 | £10 | £5 |
| **Batman Figure Assortment** <br> Batman and Robin are now ready for any assignment. 183 x 302 x various mm. | £20 | £10 | £5 |
| **Powering Batman** <br> Black and gold. Issued 1993. | £10 | £3 | £1 |
| **Air Attack Batman** <br> Black, purple and gold. Issued 1993. | £10 | £3 | £1 |
| **Batman All-Terrain Vehicle – BATV** <br> Black or dark blue with skis and decor. Versatile mini-vehicle with front skis which can convert into roller blades. Batman can fire missiles from the battle mode position with the front hood acting as a protective aid. Issued 1993. | £30 | £10 | £5 |

**Batsignal**
Black or dark blue with decals. Stealthy missile firing
vehicle camouflaged against the night sky. Issued 1993.
222 x 150 x 76mm.                                £10    £5    £2

**The Hoverbat**
Silver, black and lemon or gold. One-man flying machine
with dual manipulator arms and capture claw. Rear
docking station allows Batman to carry the enemy away.
Issued 1993.                                     £20   £10    £5

**The Batcycle 2**
Black or dark blue. Motorised turbo power and a
permanently attached Batman figure in authentic livery
and motifs, a super speed lifelike model. Issued 1993. 305
x 202 x 55mm.                                    £20   £10    £5

**Turbojet Batman**
Black or dark blue with silver tint and silver and yellow
weapon. Issued 1993. 183 x 302 x 56mm.          £20    £6    £3

**Combat Belt Batman**
Silver-grey, black and lemon with weapons and decor.
Issued 1993. 183 x 302 x 56mm.                  £20   £10    £5

**Deep Dive Batman**
Yellow suit with black headgear, mask and sea diving
equipment. Issued 1993.                          £10    £3    £1

**Laser Batman**
Purple, yellow and black. Issued 1993.           £10    £3    £1

**Bruce Wayne**
Black with gold and red decor. Issued 1993.      £10    £3    £1

**Bruce Wayne Custom Coupe**
Black with white wheels or silver, with Bruce Wayne
figure in black and purple suit. Issued 1993. Dimensions
of vehicle 206 x 95 x 432mm.                     £50   £20   £10

**Heavy Metal Cycle**
White, red and silver with large black wheels. This
amazing bike can deploy a hidden battle shield and
launch an awesome missile which can destroy any foe.
Issued 1993. 153 x 70 x 255mm.                   £30   £15   £10

**Catwoman**
All black. Issued 1993.                          £10    £3    £1

**Laser Blade Cycle**
Black with silver suit. Issued 1993.             £20    £6    £3

| KENNER PARKER | MB | MU | GC |
|---|---|---|---|

**The Joker Mobile**
The outstanding feature is the light blue face with white mouth, red lips and dark eyes. Metallic blue body with metallic silver wings, large yellow or gold spoked wheels and yellow-gold top and the Joker figure with silver face and purple jacket. A model filled with surprises as the bonnet slides back to change the car into a mini-assault vehicle tank with anti-aircraft-type guns popping out on top. Pull the exhaust pipe and the Joker's teeth actually launch. Issued 1993. 230 x 152 x 89mm. Great investment.  £50  £20  £10

**The Nerf Pocket Rocket**
Red and black with yellow fins and the word 'Nerf' in yellow-orange. This high powered pocket sized launcher blasts pocket-size foam arrows fast and far. Issued 1993. 178 x 54 x 305mm.  £20  £10  £5

**Penguin**
Black and white suit with white, black and red umbrella and backpack. Issued 1993.  £10  £3  £1

**The Penguin Figure 2**
Gold and white suit and black cape with top hat, red and white umbrella and weapons. Issued 1993. 183 x 302 x 56mm.  £30  £15  £7

**The Penguin Umbrella Jet**
Red, silver, black and white with silver plastic wheels. Issued 1993.  £30  £10  £5

**The Riddler**
Black, silver and green outfit with black and green weapons. Issued 1993. 183 x 302 x 56mm.  £20  £10  £5

**Robin**
Red, black and gold with backpack. Issued 1993.  £10  £3  £1

**Robin Figure 2**
Red jacket, green trousers and black boots with silver and orange or yellow machine and silver and black weapons. Issued 1993. 183 x 302 x 56mm.  £20  £10  £5

**Robin Jet Foil Cycle**
Black and red with yellow footpiece. Issued 1993.  £20  £6  £3

**Sky Drop Airship**
Black with purple wheels and wings and gold towbar. Issued 1993.  £30  £10  £5

**T2 Stronghold Mobile Armoury**
Black, silver and orange livery. Provides the ground
support the Terminator needs. Packed with giant assault
cannon, launching catapult and bazooka blaster. Cockpit
with opening canopy. Issued 1993. 267 x 127 x 419mm.    £30    £20    £10

**T2 Skyhammer Flying Vehicle**
Brick-red or brown and silver. Provides the Terminator
with air superiority against the Cyborgs. Flying chopper
that comes with two large bombs, manually activated to
launch. Issued 1993. 178 x 95 x 267mm.    £30    £15    £10

**Power Arm Terminator**
Black, silver and orange with missile launch and
grabbing claw. The arm is also interchangeable. Issued
1993.    £10    £5    £3

**Techno Punch Terminator**
All-silver with super smashing action, with red eyes
glowing. Techno Punch throws an awesome super punch
which can demolish any enemy. Issued 1993.    £15    £8    £4

**Exploding T1000**
Silver with blast apart action. After being hit by the
Terminator, the evil exploding T1000 completely blows
apart, but he can re-form himself to do battle again.
Issued 1993.    £15    £8    £4

**Hidden Weapon Terminator**
Black and orange with hidden chest cannon. Reveals his
hidden Cyborg endo-skeleton and shows his huge chest
cannon. Issued 1993.    £15    £8    £4

**Meltdown Terminator**
Black, gold and silver. Armed with water spraying
backpack. Large chunks of his bioflesh are missing, burnt
away by the white hot liquid he sprays. Issued 1993. 178
x 305 x 42mm.    £10    £4    £2

**White Hot T1000**
Made of gold coloured amalgam, this is the toughest
robot yet. Spray him with warm water and his body gets
covered with fissures, lines that make him weak and
powerless. Cool him down and he returns to his former
self. He is armed with a missile launching gun in white
livery. Issued 1993. 178 x 305 x 42mm.    £20    £10    £5

**Endglow Terminator**
Black and orange. His body glows in the dark and he is
armed with a flame thrower blaster, shooting flame
missiles at his foes. Issued 1993. 178 x 305 x 42mm.    £15    £7    £3

| KENNER PARKER | MB | MU | GC |
|---|---|---|---|

**Two Face**
Black and white suit with blue or silver and red weapon,
with gold chain. Issued 1993. 183 x 302 x 56mm.

| | £20 | £10 | £5 |
|---|---|---|---|

**T2 Mobile Assault Vehicle**
Black streamlined design with streaking white decor and
black rocket launchers with grey rocket. Silver and black
wheels. The ultimate battle car. Issued 1993. 190 x 94 x
293mm.

| | £40 | £20 | £10 |
|---|---|---|---|

**Legend of Zagor**
'Who Dares Challenge Him?' Over 45 detailed miniatures based on the characters from Ian Livingstone's best
selling fantasy books. Issued 1993.

| | MB | MU | GC |
|---|---|---|---|
| Unsigned | £20 | £10 | £5 |
| Signed by Ian Livingstone | £1000 | — | — |

# LEGO

Lego has been one of the most successful firms in the entire history of playthings, and became so successful that it created a tourist empire almost in line with Disneyland. If you have never been to the Legoland holiday village, you are missing a treat. You can also join the Lego Builders Club by writing for details to: Wrexham, Clwyd LL13 7TQ.

| MODEL | MB | MU | GC |
|---|---|---|---|
| **No. 305  Two Crater Plates**<br>Grey or slate. Issued 1982. Deleted 1987. | £2 | — | — |
| **No. 306  Landing Strip and Launching Pad**<br>Grey, silver or white with wide yellow rings. Issued 1982. Deleted 1988. | £5 | — | — |
| **No. 885  Space Scooter**<br>Grey, blue, red and black. Issued 1980. Deleted 1983. | £10 | £6 | £3 |
| **No. 886  Space Buggy**<br>Silver with red wheels and black tyres. White spaceman in silver and green with weapon. Issued 1982. Deleted 1990. | £30 | £10 | £5 |
| **No. 889  Radar Truck**<br>Grey or dark pink with red wheels and black tyres. Issued 1980. Deleted 1985. | £20 | £10 | £5 |
| **No. 891  Two Seater Space Scooter**<br>Grey with red tips and white side rockets. Issued 1980. Deleted 1985. | £15 | £8 | £4 |
| **No. 894  Mobile Signals Centre**<br>Light green, off-white or grey with red wheels and black tyres, green tip on scanner. Issued 1980. Deleted 1985. | £20 | £10 | £5 |
| **No. 897  Mobile Rocket Launcher**<br>Silver with red wheels and black tyres, with two spacemen, one in red suit and the other in white. Black and white rocket. Issued 1982. Deleted 1990. | £40 | £20 | £10 |
| **No. 918  One Man Space Ship**<br>Yellow, blue and grey with a dash of green on legs. Issued 1980. Deleted 1983. | £20 | £10 | £5 |

**No. 920  Rocket Launching Pad**
Silver, grey and blue with black and white rocket, spacemen and buggy. Issued 1982. Deleted 1987.

|  | £40 | £20 | £10 |
|---|---|---|---|

**No. 924  Space Transporter**
Blue, yellow or gold, grey and white with black wheels. Issued 1980. Deleted 1983.

|  | £20 | £10 | £5 |
|---|---|---|---|

**No. 928  Space Cruiser and Moonbase**
Silver-grey, blue, yellow and black livery. Issued 1982. Deleted 1990.

|  | £20 | £10 | £5 |
|---|---|---|---|

**No. 6801  Space Scooter**
Silver and black with spaceman in white suit and weapon in black. Issued 1982. Deleted 1990.

|  | £30 | £10 | £5 |
|---|---|---|---|

**No. 6821  Lunar Rock Collector**
Silver with red wheels, black tyres and red, yellow and black scoop. White spaceman. Issued 1982. Deleted 1990.

|  | £30 | £20 | £10 |
|---|---|---|---|

**No. 6822  Space Grab**
Silver and black with spaceman in red suit. Issued 1982. Deleted 1990.

|  | £30 | £20 | £10 |
|---|---|---|---|

**No. 6841  Survey Vehicle**
Silver, red and black. Issued 1982. Deleted 1990.

|  | £30 | £10 | £5 |
|---|---|---|---|

**No. 6842  Inspection Spacecraft**
White, black and silver with space figure in red suit. Issued 1982. Deleted 1986.

|  | £40 | £10 | £5 |
|---|---|---|---|

**No. 6870  Spacecraft Launcher**
White, silver and black with red wheels and red driver. Issued 1982. Deleted 1987.

|  | £30 | £10 | £5 |
|---|---|---|---|

**No. 6880  Mercury Surveyor**
Silver, black and yellow. Issued 1982. Deleted 1986.

|  | £30 | £10 | £5 |
|---|---|---|---|

**No. 6890  Delta 1 Explorer and Shuttle**
White, blue, black and red with red spaceman. Issued 1982. Deleted 1987.

|  | £40 | £20 | £10 |
|---|---|---|---|

**No. 6927  Mobile Tracking Station**
White, blue and yellow with silver or white wheels and black tyres. Silver scanners with green tips. Issued 1982. Deleted 1988.

|  | £40 | £20 | £10 |
|---|---|---|---|

**No. 6929  Space Transporter**
Silver, blue, white and black with grey-green tipped scanners and red driver. Issued 1982. Deleted 1988.

|  | £40 | £20 | £10 |
|---|---|---|---|

| LEGO | MB | MU | GC |
|---|---|---|---|

**No. 6950  X15 Satellite Launcher**
Silver and yellow with large black or dark blue wheels.
Black, white and silver rocket. Issued 1982. Deleted 1986.    £30    £20    £10

**No. 6970  Command Centre**
Grey, silver, black, blue, white, lemon and red with
scanners. Four space people each in red, silver, white and
grey. Issued 1982. Deleted 1990.    £40    £20    £10

# MATCHBOX

The famous Matchbox Company has come a long way since, in the late 1870s, Moses Kohnstam founded a toy company at Fuerth, near Nuremberg, and began to make quality toys under the trademark 'Moko', taken from the first two letters each of Moses and Kohnstam. For the full story, readers should consult my *Matchbox Toy Price Guide*.

| MODEL | MB | MU | GC |
|-------|-----|-----|-----|

### ADVENTURE 2000

**K2001  Raider Command, First Issue**
Fawn body with fawn or orange tracks, orange interior, seats with red, white and blue stripes and flag decals. Spacemen and superspace wheels. Price £3.99. Issued 1978/79. Deleted 1982. 162mm.

| | £100 | £50 | £20 |
|-------|-----|-----|-----|

**K2001  Raider Command, Second Issue**
Metallic blue body, golden orange and yellow interior seats. Black rocket gun with red rocket and spacemen. Price £3.99. Issued 1982. Deleted 1986. 162mm.

| | £100 | £50 | £20 |
|-------|-----|-----|-----|

**K2002  Fighter Hunter, First Issue**
Fawn body with gold and silver engine. Red space wings. Superspace wheels and spacemen. Price £3.50. Issued 1978/79. Deleted 1982. 117mm.

| | £100 | £50 | £20 |
|-------|-----|-----|-----|

**K2002  Fighter Hunter, Second Issue**
Metallic blue body and red wings. Silver space guns, silver and gold interior, blue-silver fin and spacemen. Price £3.75. Issued 1980. Deleted 1986. 117mm.

| | £60 | £30 | £15 |
|-------|-----|-----|-----|

**K2003  Crusader, First Issue**
Fawn body with off-white interior and orange tinted windows. Red rockets and space guns. Black tracks, superspace wheels, spacemen and decals. Price £3.99. Issued 1978/79. Deleted 1982. 113mm.

| | £100 | £50 | £20 |
|-------|-----|-----|-----|

**K2003  Crusader, Second Issue**
Metallic blue with orange dome lights and tinted windows. Red space guns and spacemen. Black tracks and superspace wheels. Design and '2000' decal on each side. Price £3.50. Issued 1980/81. Deleted 1986. 113mm.

| | £100 | £50 | £20 |
|-------|-----|-----|-----|

### K2004  Rocket Striker, First Issue
Metallic fawn body with red rocket holders and black rockets. Red scanner and radar piece. Decals, clear windows, black tracks and superspace wheels. Price £3.75. Issued 1978/79. Deleted 1982. 112mm.

| | £100 | £50 | £20 |

### K2004  Rocket Striker, Second Issue
Metallic blue with red rocket holders and rocket guns. Radar scanner, decals in black and orange, green or blue tinted windows, spacemen, black tracks and superspace wheels. Price £3.99. Issued 1980/81. 112mm.

| | £100 | £50 | £20 |

### K2005  Command Force, First Issue
Three vehicle set in metallic blue: space buggy, rocket car and hovercraft. Silver trim, lights, grilles etc. Spacemen, superspace wheels, decals and red roof lights. Price £3.50. Issued 1980/81. Deleted 1986.

| | £100 | £50 | £20 |

### K2005  Command Force, Second Issue
Deeper metallic blue with red rockets, thick black plastic tracks, superfast space wheels and shuttle launcher. Decals on sides. Silver radar scanner and guns. Price £4.50. Issued 1981. Deleted 1986. 160mm.

| | £100 | £50 | £20 |

### K2006  Shuttle Launcher, First Issue
Deep metallic blue with red rocket launcher above. Thick black plastic tracks. Superfast wheels and shuttle launcher decals on sides. Silver radar scanner and guns in silver. Price £4.50. Issued 1981. 160mm.

| | £50 | £20 | £10 |

## BATTLE KINGS

### K111  Missile Launcher, First Issue
Bright green or lime-green with orange launcher and scanner. Metal wheels with thick tyres, black and matching rocket livery. Price £2.45. Issued 1976. Deleted 1970. 112mm.

| | £50 | £25 | £10 |

### K111  Missile Launcher, Second Issue
Medium green with orange-red launcher pads and scanner. Blue tinted windows, thick black plastic wheels with black hubs. Three soldier figures in lime-green. Price £2.65. Issued 1978. Deleted 1980. 112mm.

| | £50 | £25 | £10 |

### K111  Missile Launcher, Third Issue
Camouflage two-tone dark green livery with red launching pads and scanner, black rockets and thick black plastic wheels. Price £3.25. Issued 1980/81. Deleted 1986. 112mm.

| | £40 | £20 | £10 |

### K117 SP Rocket Launcher, First Issue
Light fawn and dark green camouflage body with black
rocket launcher and red rockets. Green wheels and dark
tracks. Price £2.50. Issued 1978. Deleted 1980. 105mm.

£100 £40 £15

### K117 SP Rocket Launcher, Second Issue
Light green and mustard or dark fawn body. Black rocket
launcher pads with red rockets, green wheels and dark
tracks. Price £3.25. Issued 1980/81. Deleted 1986. 105mm.

£50 £25 £10

## CONVOY

### CY-AP204 NASA Set, First Issue
Convoy action pack consisting of large vehicle carrying a
rocket plus a van and helicopter. White, grey and orange
with NASA decals and wording. Price £14.75. Issued
1987. Deleted 1990.

£100 £50 £20

### CY NASA Set, Second Issue
Limited edition with new livery and markings. Price £15.
Issued 1988. Deleted 1990.

£150 £40 £20

### T-5 NASA Set, Third Issue
Team Convoy series. White with grey trailer. NASA
decals and wording on sides of the two vehicles. Price
£15. Issued 1988.

£50 £20 £10

## SKYBUSTERS

### SB3 NASA Space Shuttle, Fourth Issue
Brilliant white body with black jets. USA and US decals
in black on yellow-gold background. Flag decals. Price
75p. Issued 1980/81. Deleted 1987. 100mm.

£50 £25 £10

### SB3 NASA Space Shuttle, Fifth Issue
White with US flag, decals and wording. Price 99p. Issued
1988. Deleted 1992.

£20 £10 £5

### SB3 NASA Space Shuttle
Limited edition in white and black with US decals.
Decorative display box. Issued 1993. High quality model
which will soon be snapped up.

£50 £20 £10

## RING RAIDERS

This innovative boys' toy line offered an exciting adventure alternative to the
lack-lustre action figure category. The ring would fit on to the second and fourth
fingers. The various Ring Raider models would fit and be fired from the finger.

There were other gadgets made to fit on to the arm with controls, instruction book, figures and a comic book. Issued 1989. Deleted almost at once. The following is a complete list of what were made in catalogue order. Four-pack series 1.

### Set 1  Commander Victor Vector
This set of planes of the future includes the above commander in blue, red, black and gold markings and livery; F-19 plane in red, white, black and blue livery and markings; F-104 plane in black, white and red; X-29 plane in red, blue, black and white; and Harrier jet in red, white, blue and black.        £50    £30    £15

### Set 2  Cub Jones F-5 Command Plane Skytiger
Includes the above with F-5 plane in orange and black with red and white rockets; P-51 plane in yellow-gold and black; Viggen plane in mustard and black; and F-86 plane in blue-grey and white livery and logo.        £50    —    —

### Set 3  Yinsu Yakamura
Includes the X-29 command plane Samurah flyer in red and yellow with white line trim; F-19 plane in red, white and yellow-gold trim; Harrier jet in lemon or mustard and red; and F-86 plane in red and yellow livery and markings.        £50    —    —

### Set 4  Chiller
The above figure is in blue and black outfit with wide white collar. His planes include F-104 rocket shape, in white body with black and blue design, and red markings on nose; F-19 in black, white and blue with a red nose-tip; P-51 in black with blue, black and white design on wings and tail, with blue prop; and the smart X-29 in black and white with red dash.        £50    —    —

### Set 5  Viggen Command Craft
Viggen plane is half black and half yellow and flame-red design; Harrier jet in black and white with red nose and white markings; F-5 plane in black and mustard with red nose-part; and F-86 plane in black and yellow with red cockpit interior.        £50    —    —

### Set 6  Wraither
The above leader in pink and blue with red mask, with P-51 plane in black and pink with green wing, tail and nose-tips; P-5 plane in purple and black with green tips and red cockpit interior; P-104 in pink and black with green tips and red cockpit interior; and Viggen plane in pink and black with green tips and red cockpit interior.        £50    —    —

### Set 7  Joe Thundercloud (Arrowhead)
With the F-86 command plane in white, red and blue designs; F-4 plane with blue, red, white and yellow designs; Corsair with its white tail and gaily multicoloured body and black prop; MIG-29 with white nose and blue rear, with golden design on wings.     £50    —    —

### Set 8  Hubbab Zap-Master
With the MIR-111 command plane in black and yellow lined livery and red nose; Corsair in black, white and red, with blue nose and black prop; F-4 plane in black, orange and red design; and A-10 plane in purple and black with gold lines and red nose.     £50    —    —

### Set 9  Max Miles (Air Knight)
With the SR-71 command plane in dark blue, yellow or gold, and white with red tipped rockets; F-16 plane in red, white, gold and dark blue; Mirage 1 in nearly all-white and blue dots with badge; F-19A plane in mainly green with white nose-tip and decals.     £50    —    —

### Set 10  Black Jack (Reckless Raven)
With the Harrier command plane in black with orange edged markings and white and blue design; Corsair in black with yellow and blue edging black prop; A-10 plane in mainly black with golden yellow edging and red cockpit interior; F-19A plane in black, orange, white and red with white design.     £50    —    —

### Set 11  Yuri-Kirkov (Kirkov's Comet)
With the F-4 command plane in yellow, orange-red and white with blue cockpit interior; Mirage 111 in identical livery as F-4; SR-71 plane in matching livery of orange, yellow and white; A-10 plane in yellow, red and white with blue cockpit interior.     £50    —    —

### Set 12  Mako (Sea Harrier)
With the MIG-29 command plane in black, dark blue, red, white and light blue livery; F-19A plane in black and dark yellow with red cockpit interior; F-16 plane in blue-green, black and red; and SR-71 plane in black with black body and pink crab-like design on top.     £50    —    —

## POCKET ROCKETS

Smart future motorbikes which must be in the world of space, speed and action.

**The Four Motorbike Set**
White, red, black and silver with famous name and
decals. Price for each £2.25. Issued 1988. Deleted 1989.

| | | | |
|---|---|---|---|
| Single model | £50 | £20 | £10 |
| Set of four | £250 | — | — |

**Colani Racers Set**
These specially designed racers are definitely out of this
world. Luigi Colani, the world-famous designer, created
these superb futuristic cars complete with high speed
motors, which were meant for the high quality,
quick-thinking collector. Each has its own livery: metallic
blue, silver, red and yellow colours with silver wheels,
hubs, lights, trim and black tyres. Price for each £3.75.
Issued 1988. Deleted almost at once.

| | | | |
|---|---|---|---|
| Single model | £50 | £25 | £10 |
| Set of four | £250 | — | — |

**Pocket Rocket Buggies**
Another set of four racers, Miniature Hotrods, with full
working suspension, that really scorch along. Decorated
in bright livery to add extra colour excitement: dark blue,
fawn and black (rare and worth treble). Price for each
£3.25. Issued and deleted 1988.

| | | | |
|---|---|---|---|
| Single model | £40 | £20 | £10 |
| Set of four | £175 | — | — |

## 1–75 SERIES

**No. 40  Rocket Transporter**
Unusual model in white or grey with US decals and space
designs. Price 99p. Issued 1986. Deleted 1990. 75mm.

| | £10 | £5 | £1 |
|---|---|---|---|

**MB Miniature 40  Rocket Transporter**
White with red markings and wording, with silver
wheels and black tyres. Price £1.10. Issued 1989. Still
current 1993.

| | £10 | £5 | £1 |
|---|---|---|---|

**No. 49  Dune Man Special**
Red or orange with silver engine, black bumpers and
tyres. Dune Man motif and wording on sides and bonnet.
Price 99p. Issued 1986. Deleted 1987. 65mm.

| | £20 | £10 | £5 |
|---|---|---|---|

**No. 54  NASA Tracking Vehicle**
White or cream with brown or purple stripe around
centre. Orange interior. Superfast wheels, black grille and
bumpers. The words 'Space Shuttle Command Centre' on
sides and 'NASA Tracking Station' on roof in thick blue
lettering. Also red and blue stars on roof and NASA in
thick red letters on roof. Tracking system on roof. Price
64p. Issued 1982. Deleted 1988. 72mm.

£30    £15    £5

**MB Miniature 57  Mission Helicopter, First Issue**
White and red with blue tinted windows. Price £1.10.
Issued 1989. Deleted 1992.

£15    £5    £2

**MB Miniature 57  Mission Helicopter, Second Issue**
White, red and black tracks. New design for 1993.

£10    £5    £2

**No. 59  Planet Scout Car**
Rich green tone on top half of body and bright yellow on
lower part, including base. Silver headlights, bumpers
and grille; yellow bumpers found on some models.
Models have orange or yellow interior, or off-white
interior and orange tinted windows. Wide superfast
wheels. Price 64p. Issued 1976. Deleted 1979. 70mm.

£30    £20    £10

**No. 67  Hot Rocket Special**
Metallic lime-green livery with silver motor, grille,
headlights and bumpers. Superfast wheels. Price 15p.
Issued 1974. Deleted 1979. 76mm.

£30    £20    £10

**Matchbox Corky PX-103  Corky's Starship**
This was the only space outfit and Great Adventure
cassette tape. Silver space suit with red shoes, Corky and
US flag logos. Red, white and blue. Issued 1988. Deleted
almost at once.

£50    £25    £10

**GF310  Monster Truck**
A collector's piece which must be unused to retain its
highest value in years to come, but I advise any person
who can afford it to buy two. Then you can build and use
one as it was meant to be. You can design your own
personal monster-crushing truck, complete with its own
highly detailed crushed base. Advertised in the catalogue
as white with large wheels. Issued 1993. 250 x 50 x
280mm.

£30    £10    £5

**Toddler Toys LL-2440  Space Station**
Pre-schoolers can pretend that they are in outer space
with this 'out of this world' playset. Red, white, yellow,
black and blue. The control tower has a work centre and
a crank operated see- through elevator with radar discs
that also rotate. The child can explore in the 8-wheel

commander van or take the space-ranger truck complete with camera to snap the two strange space creatures in their two-legged space ship. Issued 1988. Deleted 1992.

£100 £50 £20

### PK-5102 Star Trek USS Enterprise (also numbered PK-5110)
Featured in the Star Trek TV series, Captain Kirk's flagship is used for Intergalactic exploration missions. Price £5.75. Issued 1979. Deleted almost at once. Limited production.

£100 —

### PK-5105 Star Trek K-7 Station
This complex kit serves as a combined space trading station and recreation centre. Price £5.75. Issued 1979. Deleted almost at once. Limited production.

£100 — —

### PK-5106 Star Trek Romulan Space Ship
This is the model of the Romulans battle ship, painted like a giant bird of prey. Price £5.25. Issued 1979.

£100 — —

### PK-5110 Star Trek USS Enterprise
New size for one of the most famous names in the world. More than 22in. long, coming complete with working lights, Star Fleet decals, and a display stand to keep model firm or for exhibitions. Price £10.95. Issued 1980. Deleted 1986.

£200 — —

### PK-5111 Star Trek Klingon Battle Cruiser
The bitter enemy of Star Fleet's USS Enterprise, this is the attack vehicle of the Klingon Battle Fleet. The kit builds to about the same size as the Enterprise. Price £6.35. Issued 1980. Deleted 1986.

£100 — —

### PK-5112 Star Trek Shuttle Craft
Consists of two parts. The bottom section is a sledge, and the top is a detachable personnel carrier which transports Spock and other to and from the Enterprise. Price £5.45. Issued 1980. Deleted 1986.

£100 — —

### RG-8120 Air Award Series Medals
Six-medal set with two planes only available with this set, in special decorations, awarded only to the bravest of Ring Raider pilots. Medals include Leadership, Speed, Cool Head, Justice, Ready and Dog Fight. Complete with free special comic.

£250 — —

**ASSTD 8130  Battle Blaster Sound Machine**
This electronic machine brings realistic action sounds to all Ring Raider planes and features: machine gun bursts; bomb explosions; laser cannon blasts, which with its unique two-position design works on the wrist or on the table. Includes adjustable velcro wrist band. Available in black camouflage for Skull Squadron and beige camouflage for Ring Raiders. Each set comes with one battle-ready plane that is possibly rare and not available with normal sale planes.     £150     —

**RG-8130  Sky Base Courage**
Fully operational Ring Raiders Sky Base with landing and take-off runways, control tower, antenna and gun turrets. Complete with one special plane not available in the normal sets. Model comes in attractive livery, decals and markings of blue, black, yellow and orange, and white. Colours can vary and make prices higher, so colours should be checked with an expert. A good investment that can be joined up with Sky Base Freedom.     £250     —     —

**RG-8140  Sky Base Freedom**
Additional Ring Raiders Sky Base can be interlocked with Courage Sky Base to form a huge aircraft base, with extended runway and hanger bays, complete with one special plane not otherwise available. This base is in orange, fawn and black design with decor on the outside with decals.     £100     —     —

**RG-8160  Skull Squadron Mobile Base**
Evil Skull Squadron Mobile Base flying headquarters complete with landing pad runway, control tower and aircraft hangers. Includes one Skull Squadron plane not available in normal sales items. Rare. Black, fawn or pink, yellow, blue and red with markings, although colours can vary.     £150     —     —

**RG-8170  Ring of Fire Video**
This action packed 30-minute VHS cassette is a cartoon mini-feature film.     —     £50     —

**RG-8180  Air Carrier Justice**
Flagship of the Ring Raiders team which transports pilots and planes through time and space. Complete with its own hangers, runways, attack and defence positions, and storage space for at least 24 planes. Doubles up as a Ring Raiders carry-case complete with its own planes which are possibly not available with normal sales items. Good investment.     £250     —     —

**RG-8190 Wing Command Display Stand**
Purpose built display stand to hold the Ring Raiders
planes when not on a mission. Complete with 2 special
planes not available in other sets. Holds up to 12 planes.
Blue stand with design. Price with 12 planes includes the
2 special planes.                                                                £250     —      —

## THUNDERBIRDS (A Very Special World)

At long last, 'Thunderbirds are Go' and some of the most sought after investment
toys ever made. In 1993 the models were introduced once more, so join in the
success of International Rescue and the hottest collectables of the year. Issued
1993. Deleted almost at once.

**TB1  Scott Tracey**
Grey-black and blue with red nose-tip.                            £50      —      —

**TB2  Virgil Tracey**
Green with yellow and orange-red markings.                  £50      —      —

**TB3  Alan Tracey**
Mostly red with black and white markings.                     £50      —      —

**TB5  Lady Penelope's FAB1**
Pink,  silver and white. Good investment.                        £100    —      —

**TB700  Thunderbirds Rescue Pack**
Complete gift set containing the entire diecast range,
Thunderbirds 1, 2, 3, 4 and 5. The high quality window
box includes biographies of the team.                             £350    —      —

**TB710  Tracey Island Electronic Playset**
Open the hanger door and drive through to Thunderbird
2. Tilt the ramp (palm trees automatically fold down), lift
the blast shield and Thunderbird 2 is ready to launch.
Slide back the swimmingpool to reveal Thunderbird 1's
secret launching silo. Thunderbird 3 launches through
the round house. A selection of four pilot commands and
rocket sounds of the Thunderbird craft. 515 x 210 x
345mm.                                                                           £250    —      —

**TB720  Thunderbird Playset 2**
This is Thunderbird 2. 'I've arrived at the scene of the
disaster. Now Virgil can fly to the rescue, lower the pod
and launch Thunderbird 4.' With three sounds, including
voice commands and rocket effects. Landing legs and
loading cockpit.                                                                £250    —      —

**TB750  Thunderbird Figures**
Highly detailed articulated figures include all the
Thunderbird characters in authentic livery, complete
with individual accessories: Parker, John, Virgil, Jeff,
Brains, Lady Penelope, Hood, Alan, Scott and Gordon.
Each figure 95mm. Set.  £500  —  —

**TB790  Thunderbirds Pullback Action Vehicles**
With high powered windback motor.  £150  —  —

STINGRAYS

**SR200  Stingray and Terrorfish**
Captain Troy Tempest's Stingray submarine and the
enemy's dangerous Terrorfish, a highly detailed craft in
authentic decals and livery as seen on TV.  £200  —  —

**SR210  Marineville**
Headquarters of the Stingray team. It lowers into its base
when under attack and defends itself with its high
performance Wasp rockets. Authentic livery and decals.
Well boxed.  £150  —  —

**SR220  Stingray Action Playset**
Highly detailed action vehicle fully compatible with
Matchbox Stingray figures. Silver, blue and lemon with
authentic markings etc. Lift-off cabin roof to load figures.
Interior seats two figures to pilot crafts. Four torpedoes
can be fired with launch mechanism. Periscope slides up
and down, and rear turbine rotates by hand. 400 x 150 x
200mm.  £100  £50  £20

**SR250  Stingray Figures**
Articulated action figures in authentic livery:
Commander Shore; Tita, leader of the Aquaphibians; and
Marina. Each figure 95mm. Set.  £200  —  —

**SR251  Captain Troy Tempest**
95mm. Very good investment.  £50  —  —

PET MONSTERS (Out of This World)

In the early part of the 1990s Matchbox brought out this series to end the 2000
series. Rare, although still to be purchased at a reasonable price. Good
investments for the young.

**SH100  Moon Shoes**
Matchbox called these playful 'out of this world,
planet-like' moon shoes 'kid powered anti-gravity'

devices. They had special adjustable bands and straps to
secure the shoes and to let kids jump and leap about with
a 'floating through space' action. The shoe bases have a
non-skid attachment. Orange, blue, black and red. Issued
1991. Deleted almost at once. Each shoe 750 x 275 x
350mm.

|  | £100 | — | — |
| --- | --- | --- | --- |

### PM2600  My Pet Monster
Beast from the planets of rough and tumble fun. Soft
enough to hug and just scary enough to keep other
monsters away. Blue, pink and orange. Vinyl handcuffs
with breakaway chains. White fang teeth, blue nose and
claw hands and feet in light brownish silver with blue or
green toes and claw tips. 26in. (660mm) high.

|  | £100 | — | — |
| --- | --- | --- | --- |

### PM2720  Monster Pet Assortment
There are three more in this set. Made of colourful
durable plush and vinyl, they have puppet action mouths
and vinyl handcuffs with breakaway chains. Colours
vary in pink, white, red, blue, orange and purple. Claw
hands and feet and tusky teeth. Each figure 12in. (305mm)
high.

| Individual model | £50 | — | — |
| --- | --- | --- | --- |
| Set of three | £150 | — | — |

### PM2802  My Pet Monster Junior
There are three varieties in this set. All have breakaway
handcuff chains. Colours vary in blue, pink, red, orange,
purple, pink and white. Claw hands and feet and tusky
teeth. Each figure 18in. (460mm) high.

| Individual model | £50 | — | — |
| --- | --- | --- | --- |
| Set of three | £200 | — | — |

# MB

MB has been a household name for many years, especially with its games and those many pastime toys for children and grown-ups alike. Yes, they are collectable, although to get the best investment returns they should not have been opened or played with. This is a time when one should buy two items, one to keep as an investment and the other to play with. Space toys and games are items that will always interest the genuine collector, and the following are some investments to look out for.

| GAME | MB | MU | GC |
|---|---|---|---|
| **Battle Masters**<br>The epic game of fantasy battles on an immense scale. The ultimate adventure with over 100 highly detailed miniatures and an immense battle mat. Issued 1993. 153 x 137cm. | Current price | | |
| **Battle Masters, Reinforcements**<br>Throughout the Empire the call has gone out to swell the ranks of the Imperial Army. Contains 26 detailed miniatures. Issued 1993. 352 x 278 x 50mm. | Current price | | |
| **Battle Masters, Chaos Warband**<br>The Army of Chaos has again been summoned to its foul banner. Contains 24 detailed miniatures. Issued 1993. 352 x 278 x 50mm. | Current price | | |
| **Hero Quest, Advanced Quest**<br>Deep inside another dimension face battling. Includes 12 miniatures. Issued 1993. 515 x 320 x 70mm. | Current price | | |
| **Hero Quest, Wizards of Morcar**<br>The Minions of Morcar have armed themselves with an evil and terrible magic. With spell cards. Contains 16 figures, 24 weapons and 64 cards. Issued 1993. 225 x 320 x 50mm. | Current price | | |
| **Hero Quest, Against the Ogre Horde**<br>A huge expansion pack with 12 new monster figures. Issued 1993. 225 x 320 x 50mm. | Current price | | |
| **Hero Quest, Kellar's Keep**<br>The legend grows with 15 more figures, new colour gameboard and 10 exciting adventures. Issued 1993. 175 x 225 x 50mm. | Current price | | |

### Hero Quest, Return of the Witchlord
The legend still grows with 16 more figures and new
gameboard sections. Issued 1993. 175 x 225 x 50mm.

Current price

### Hero Quest, Dungeon Design Kit
Hero Quest collectors can now create their own quests,
complete with decal sheets, enlarged character sheets and
great ideas. Issued 1993. 175 x 225 x 25mm.

Current price

### Space Crusade
A heroic battle against Alien Monsters with over 50 finely
detailed miniatures and 30 optional weapons. Issued
1993. 505 x 320 x 65mm.

Current price

### Space Crusade, Eldar Attack
A new race joins the Space Crusade. New to collect: 10
high quality miniatures, 15 weapons and 48 cards. Issued
1993. 225 x 320 x 50mm.

Current price

### Space Crusade, Mission Dreadnought
A huge expansion and gameboards with 12 new Alien
Monsters and more hideous than ever before with
increased weapon capacity. Issued 1993. 225 x 320 x
65mm.

Current price

# PALITOY

The name of Palitoy has been associated with toymaking for many years and its contribution to space models stood out in the 1980s. I would like to thank the manager of Kenner Parker of Leicester for help in my research.

| MODEL | MB | MU | GC |
|---|---|---|---|
| **Action Pack Set**<br>Cobra can launch an aerial attack on Action Force, Cobra and the CLAW. Silver-blue and red or orange. Issued 1985. Deleted 1990. | £50 | £30 | £20 |
| **AT-AT: All-Terrain Armoured Transporter**<br>The most famous weapon and vehicle of the Empire. Its jointed hips, knees and ankles allow it to cover all types of terrain. Lever controls the up, down and sideways movements of the head. The cockpit can hold two action figures. Issued 1984/85. Deleted 1990. 17½in. high x 22in. long. | £200 | £100 | £50 |
| **Radar Laser Cannon**<br>Swivels and clicks when turned. Also explodes into four pieces. In Star Wars livery. Issued 1984/85. Deleted 1990. | £35 | £25 | £10 |
| **The Tripod Laser Cannon**<br>'Energiser' unit and hose. Has movable tripods and the cannon makes a machine gun sound when turned. Issued 1985. Deleted 1990. | £50 | £25 | £10 |
| **Vehicle Maintenance Energiser**<br>Eight tools and 2 'energiser' hoses which are attachable to vehicles. Issued 1985. Deleted 1990. | £35 | £20 | £10 |
| **A-Winged Fighter**<br>Replica of the rebel vehicle from the film Return of the Jedi. A vehicle built for any battle with the Empire. Rotating side cannons (laser with clicking sound). Also features manually controlled canopy and retractable landing gear. Cockpit accommodates one Star Wars figure in red suit. Issued 1985. Deleted 1990. | £100 | £50 | £20 |

### B-Winged Fighter Vehicle

One of the most exciting models in the Star Wars collection. Remote controlled wings, gravity controlled cockpit with opening canopy and laser sound. An action figure can sit in the cockpit ready to make use of the remote landing gear when returning back to base after a dare-devil raid. Issued 1985. Deleted 1990.

£100 £50 £20

### Battle-Damaged X-Wing Fighter

This is a Luke Skywalker X-wing fighter, fresh from doing battle with the Empire. Push the simulated Artoo Deetoo and the wings open to attack position. Also has movable front landing gear and cockpit that opens to hold action figure. Issued 1985. Deleted 1990.

£100 £50 £20

### The Triad Fighter

Action Force 3-section spacecraft consisting of mother ship and two detachable wing ships. Used for high-speed search and attack missions in outer space. Mother ship's wings rotate. Model includes Captain figure in silver with red and blue stripes and decals. Issued 1985. Deleted 1990.

£75 £40 £20

### Y-Winged Fighter Vehicle

Replica of the addition to the fighting Rebel Force, as seen in the film Return of the Jedi. Cockpit takes Star Wars figure, and button activates the battery operated laser cannons and sounds. Features manually activated bomb, retractable landing pads and compartment for Artoo Deetoo. Issued 1985. Deleted 1990. Sound investment.

£150 £75 £40

### The Imperial Shuttle

Authentic Star Wars livery and decals. Model can have the wings manually lowered. Lower the movable loading ramp and recreate the Emperor's ominous arrival at the Death Star battle station, as in the film Return of the Jedi. The cockpit is big enough to hold two Star Wars figures. Take-off is easy as you control the retractable rear landing gear. Best feature of all with this model, the side of the vehicle pops off for easy access to the interior. Issued 1985. Deleted 1990. 530 x 414 x 380mm.

£100 £50 £20

### Imperial Side Gunner Ship

Replica of the vehicle used by the Empire to attack the Rebels. This great ship accommodates two action figures, bought separately as with all models. The side car swivels round 180° horizontally and 360° vertically, and is armed with clicking laser cannon. Issued 1985. Deleted 1990.

£150 £90 £40

| | MB | MU | GC |
|---|---|---|---|
| **Air to Land Interceptor**<br>Replica of the Empire's vehicle that defended the Death Star in the film Return of the Jedi. Has an opening cockpit and is well-equipped for any battle with side cannons that pop out and wings that automatically drop to attack position when button is activated. Issued 1984/85. Deleted 1990. | £150 | £75 | £40 |
| **The Interceptor Vehicle**<br>In authentic Star Wars livery and markings, one of the Empire's fastest fighter vehicles. The Tie Intercept is a high-quality design and includes a hatch that opens to hold an Imperial fighter pilot or other action figure. Press button to eject the wings and simulate battle damage and battery operated laser-cannon sound. Issued 1985. Deleted 1990. | £200 | £100 | £50 |
| **The Millennium Falcon Vehicle**<br>This play environment is packed with many of the same action features found in Han Solo's Millennium Falcon, and the cockpit canopy opens to reveal room for two action figures. The radar dish swivels and the rear deck panel lifts off to give access to an interior play environment. The Star Wars action figure of your choice fits into the gun turret chair. Secret floor panel lifts to uncover a hidden compartment. Issued 1985. Deleted 1990. | £250 | £100 | £50 |
| **One Man Sand Skimmer**<br>Bronze or gold as seen in many of the Star Wars adventures. Issued 1985. Deleted 1990. | £20 | £10 | £5 |
| **One Man Security Scout Vehicle**<br>Grey and green with blue arm holds and black or dark green tracks. Issued 1985. Deleted 1990. | £20 | £10 | £5 |
| **One Man Sniper Vehicle**<br>Black and blue. Clever and attractive model from the Star Wars collection. Issued 1985. Deleted 1990. | £20 | £10 | £5 |
| **Skystriker (Combat Jet)**<br>Dangerous mission fighter from the Star Wars series in authentic silver livery and decals. This supersonic jet fighter has pilot with parachute and six deadly missiles. The wings sweep back for the supersonic flight with retracting undercarriage. The wings sweep forward and the undercarriage lowers. Issued 1985. Deleted 1990. | £50 | £25 | £10 |

**Rebel Armoured Snowspeeder Vehicle**
The top vehicle that helped defend the secret Rebel base
on Hoth. Canopy opens to reveal a cockpit which can
hold two action figures. Issued 1984/85. Deleted 1990.   £100   £50   £25

## STAR WARS FIGURES

There were 74 character figures issued in the Star Wars collectables between 1984
and 1985; they were all deleted by 1990. Those who have a full collection have
a very good investment indeed.

**No. 1  Rebel Commander**
White or cream uniform, brown boots, peaked cap and
the figure holding a rifle in the right hand.   £10   £5   £2

**No. 2  Luke Skywalker (Hoth Battle Gear)**
White uniform, black boots and rifle, plus headgear.   £10   £5   £2

**No. 3  Prune Face**
Fawn or light brown outfit, black boots and rifle.   £6   £3   £1

**No. 4  Lobot**
Fawn or purple outfit, shoes, puffed sleeves and rifle.   £10   £5   £2

**No. 5  An Emperor's Royal Guard**
All-red cape and long rifle, black face mask.   £10   £5   £2

**No. 6  Yoda the Jedi Master**
Green body and face, wearing grey cloak with space gun.   £15   £9   £5

**No. 7  Two-Onebee (Z-1B)**
Blue and silver robot, part blue figure.   £15   £9   £5

**No. 8  The Emperor**
White face with blue cloaked outfit and hand probe stick.   £10   £5   £2

**No. 9  Luke Skywalker (Bespin Fatigues)**
Fawn uniform, black belt, brown boots and yellow stick.   £10   £5   £2

**No. 10  IG-88**
Silver with blue weapon.   £10   £5   £2

**No. 11  Ugnought**
Grey, black and blue dwarf-like figure.   £10   £5   £2

**No. 12  Lando Calrissian**
Grey-silver and black with weapon.   £10   £5   £2

**No. 13  Chewbacca**
Brown with weapon and sash.   £15   £6   £3

**No. 14  Imperial Stormtrooper (Hoth Battle Gear)**
All-white suit, black or blue weapon.  £15  £6  £3

**No. 15  Gamorrean Guard**
Fat, thickset figure in brown and grey.  £10  £5  £2

**No. 16  Imperial Commander**
Elegant figure dressed in black suit, peaked cap, belt and
gloves with weapon.  £15  £6  £3

**No. 17  Han Solo**
Greyish purple and green trench coat and weapon,
brown shoes and grey scarf.  £10  £5  £2

**No. 18  Logray the Medicine Man**
Grey and grown striped outfit with large black fork and
weapon with shoulder strap.  £15  £6  £3

**No. 19  Han Solo (Hoth Battle Gear)**
Blue jacket, matching hood with fur surround, fawn
trousers and weapon.  £10  £5  £2

**No. 20  Rebel Soldier**
White or grey suit with brown life-jacket and matching
boots with weapon and peaked cap.  £15  £6  £3

**No. 21  Luke Skywalker**
Black outfit and grey cape and green weapon.  £10  £5  £2

**No. 22  Death Star Droid**
Silver and black. Rare.  £20  £10  £5

**No. 23  Imperial Tie Fighter Pilot**
In space air pilot outfit with helmet and mask in blue or
black (worth double), belt and weapon.  £20  £10  £5

**No. 24  Ben (Obi Wan) Kenobi**
Blue headband, orange-brown cloak, suit and weapon.  £10  £5  £2

**No. 25  Jawa**
Small figure in all-black cloak and hood. Claw hands.  £15  £6  £3

**No. 26  Luke Skywalker**
White jacket with wide blue pockets, fawn or orange
leggings and weapon.  £10  £5  £2

**No. 27  Leia Boushh**
Tall, bear-faced figure with grey shorts and boots, blue
top and matching hat, and laser weapon.  £20  £10  £5

| PALITOY | MB | MU | GC |
|---|---|---|---|

**No. 28  Biker Scout**
Elegant figure in black and white with gauntlets, boots and weapon.

| | £15 | £6 | £3 |

**No. 29  Bossk (Bounty Hunter)**
Lemon, black and red neckband, gauntlets and weapon.

| | £15 | £6 | £3 |

**No. 30  Princess Leia Organa**
Green camouflage combat, poncho with black belt, blue trousers, black leggings and white hat with green band.

| | £20 | £10 | £5 |

**No. 31  AT-AT Commander**
Mainly blue or purple outfit, black boots and weapon.

| | £10 | £5 | £2 |

**No. 32  Nikto**
White, blue, grey and dark blue outfit, brown mask, gloves and weapon, belt and pack.

| | £15 | £6 | £3 |

**No. 33  Chief Chirpa**
Smaller figure in grey and chocolate, black belt and weapon.

| | £10 | £5 | £2 |

**No. 34  RS-04**
Size as No. 70. Silver with red and blue markings. Great favourite with collectors.

| | £20 | £10 | £5 |

**No. 35  C-3PO**
All-silver figure – a winner all the way.

| | £20 | £10 | £5 |

**No. 36  Lumat (Ewak Warrior)**
Small figure in brown and white with weapon.

| | £10 | £5 | £2 |

**No. 37  Wicket W. Warrick**
Very tiny figure in brown and orange with spear.

| | £15 | £6 | £3 |

**No. 38  FX-7**
Unusual, tubular figure in blue with white stick probes from all over the body.

| | £20 | £10 | £5 |

**No. 39  Paploo**
Grey and grey with probe weapon.

| | £10 | £5 | £2 |

**No. 40  Stormtrooper**
Brilliant white livery and black gloves, pockets and blue weapon.

| | £15 | £6 | £3 |

**No. 41  Rebel Commando**
Attractive figure in green with grey weapon.

| | £20 | £10 | £5 |

**No. 42  Klaatu**
Blue and grey with weapon, black boots.

| | £10 | £5 | £2 |

| PALITOY | MB | MU | GC |
|---|---|---|---|

**No. 43  Teebo**
All blue, two-tone with fawn belt.

| | £10 | £5 | £2 |

**No. 44  Boba Fett**
Blue, green, black and brown with weapon.

| | £15 | £6 | £3 |

**No. 45  Zuckuss**
All blue with weapon.

| | £20 | £10 | £5 |

**No. 46  Dengar**
Blue with white harness and weapon.

| | £10 | £5 | £2 |

**No. 47  Star Destroyer (Commander)**
Grey-blue with helmet and weapon, dark boots.

| | £10 | £5 | £2 |

**No. 48  Tusken Raider's Sand People**
Grey-white or fawn with black strap and belt, weapon and face mask.

| | £15 | £6 | £3 |

**No. 49  AT-AT Driver**
Blue-grey suit with black boots, belt and gloves.

| | £10 | £5 | £2 |

**No. 50  Han Solo**
Wearing a Bespin outfit consisting of a dark blue jacket, light blue shirt, brown trousers, fawn or orange tinted shoes, belt and weapon.

| | £10 | £5 | £2 |

**No. 51  8D8**
Very unusual figure in lemon with green hood and brown fork-weapon. Rare livery worth treble that of white and blue.

| | £10 | £5 | £2 |

**No. 52  Darth Vadar**
Black with red stick-weapon and black hood with eye-slits.

| | £15 | £6 | £3 |

**No. 53  B-Wing Pilot**
In red suit with earphones, backpack and brown boots.

| | £20 | £10 | £5 |

**No. 54  Lando Calrissian**
Purple, blue and black with purple cloak.

| | £10 | £5 | £2 |

**No. 55  Klaatu**
In a Skiff Guard outfit in off-white or grey and brown boots, belt, pouch and backstrap.

| | £20 | £10 | £5 |

**No. 56  Princess Leia Organa**
Wearing a cherry-red Bespin gown with white cloak, fur collar, necklet and blue weapon.

| | £20 | £10 | £5 |

**No. 57  Princess Leia Organa**
Another super outfit in white with matching cape, white
shoes and black laser weapon.                          £20    £10    £5

**No. 58  Han Solo**
Looking more like a cowboy from the Wild West in dark
jacket and trousers with blue shirt and weapons.       £15    £6    £3

**No. 59  Luke Skywalker**
This time as an X-wing fighter pilot with red or orange
suit, white helmet, black and white backpack, boots and
weapon.                                                £20    £10    £5

**No. 60  AT-AT Driver**
Light silver-blue or off-white suit and boots, helmet and
weapon and backpack.                                   £20    £10    £5

**No. 61  Amanaman**
Unusual figure in lemon or gold outfit with green hood
and brown or fawn fork-stick.                          £20    £10    £5

**No. 62  Imperial Dignitary**
Wearing a long blue or purple cloak outfit with a
matching hood and red blouse.                          £15    £6    £3

**No. 63  Lando Calrissian**
Silver with a flowing, long blue cape and long boots.  £20    £10    £5

**No. 64  Barada**
Lemon, grey and brown with black boots and weapon.     £10    £5    £2

**No. 65  Luke Skywalker**
This time this famous figure is wearing a poncho in grey
and light brown, black leggings, belt and headgear. Also
with weapon.                                           £10    £5    £2

**No. 66  Han Solo**
Unusual silver-blue figure in dark carbonite chamber.  £25    £15    £7

**No. 67  A-Wing Pilot**
Green flying suit with backpack and boots, helmet,
goggles and earphones.                                 £20    £10    £5

**No. 68  Imperial Gunner**
All-black livery with helmet and goggles, belt and boots.
First-class investment.                                £20    £10    £5

**No. 69  Luke Skywalker**
All-white with black gloves, white boots and weapon.   £20    £10    £5

| | MB | MU | GC |
|---|---|---|---|
| **No. 70  R2-D2 with Light Sabre**<br>This favourite figure in the range was soon snapped up when it became available. White and silver with authentic markings, red, white and blue. | £30 | £15 | £10 |
| **No. 71  Anakin Skywalker**<br>Grey suit with a long cape-like robe in black or dark blue with matching belt. | £10 | £5 | £2 |
| **No. 72  Yak Face**<br>Weird figure in white, fawn, blue and white with long spear-like missile. | £20 | £10 | £5 |
| **No. 73  Warak**<br>Small favourite in grey with dark brown hood. | £10 | £5 | £2 |
| **No. 74  Rokba**<br>Chocolate with orange spear and dark head. | £10 | £5 | £2 |

# SPACE TOYS FOR THE NURSERY AND BEYOND

Collecting can begin even before a child is born and, when one gets the urge to collect, it is at a very early age, as I know from experience in dealing with children of all ages. Just mention Bucky O'Hare, Dead-Eye Duck, Willy de Witt, Captain Planet or the Toxic Crusaders to a child for example, and you will hear plenty about the countless space heroes.

| MODEL | MB | MU | GC |
|-------|----|----|----|

## MATCHBOX

MONSTER IN MY POCKET
Selection of mini-collectable monsters, taken from myths, folklore, fables and space and the planets. There were 48 models to collect in this first series. Issued 1991.

**Baby Yaga**
Red.                                          Current price

**The Beast**
Green.                                        Current price

**Behemoth**
Purple.                                       Current price

**Bigfoot**
Purple.                                       Current price

**Catoblepas**
Green.                                        Current price

**Cerebus**
Purple.                                       Current price

**Charon**
Red.                                          Current price

**Chimera**
Purple.                                       Current price

**Coatlicue**
Yellow.                                       Current price

**Cockatrice**
Green.                                        Current price

**Cyclops**
Yellow.                                       Current price

**Ghost**
Yellow.                                       Current price

**The Ghoul**
Green.            Current price

**Goblin**
Yellow.            Current price

**Great Beast**
Purple.            Current price

**The Gremlin**
Green.            Current price

**Griffin**
Green.            Current price

**Haniver**
Yellow.            Current price

**Harpy**
Purple.            Current price

**Hobgoblin**
Red.            Current price

**The Hunchback**
Red.            Current price

**Hydra**
Green.            Current price

**The Invisible Man**
Red.            Current price

**Jotun Troll**
Purple.            Current price

**Kali**
Green.            Current price

**Karnak**
Purple.            Current price

**Kraken**
Yellow or gold.            Current price

**Mad Scientist**
Yellow.            Current price

**Manticore**
Yellow.            Current price

**Medusa**
Green.                                    Current price

**The Monster**
Green.                                    Current price

**The Mummy**
Yellow.                                   Current price

**The Ogre**
Red.                                      Current price

**The Phantom**
Purple.                                   Current price

**Redcap**
Red.                                      Current price

**Roc**
Red.                                      Current price

**Skeleton**
Yellow.                                   Current price

**Spring Heeled Jack**
Green.                                    Current price

**Tengu**
Purple.                                   Current price

**Triton**
Red.                                      Current price

**Tyrannosaurus Rex**
Purple.                                   Current price

**The Vampire**
Green.                                    Current price

**Vampiress**
Purple.                                   Current price

**The Werewolf**
Red.                                      Current price

**Windigo**
Red.                                      Current price

**Winged Panther**
Purple.                                   Current price

**The Witch**
Purple.                                          Current price

**Zombie**
Yellow.                                          Current price

**Stingray**
Authentic livery and motifs. Submarine with dual firing
sting missiles, detachable cockpit and rotating
propulsion unit. Price £29.99. Issued 1992/93. First-class
investment.                                      Current price

**Thunderbirds Playset**
Authentic Thunderbirds livery and decals with pilot
commands and rocket sounds. Retractable legs for
touchdown and take-off. Removable pod contains
Thunderbird 4 with working parts. Price £46.95. Issued
1992/93.                                         Current price

**MT-242  Secret Single Pack**
Second series. Variety of liveries. Issued 1993.   Current price

**MT-250  Twelve-Pack Super Scary**
Second series. Variety of liveries. Issued 1993.   Current price

**MT-260  Six-Pack Super Scary**
Second series. Variety of liveries. Issued 1993.   Current price

**MT-290  Cauldron Pack**
Includes a Monster Pouch for all the series; 12 monsters,
6 with special glow-in-the-dark bodies; Battle Card
Game; and Monster Cap. Issued 1993.             Current price

**MT-310  Six-Pack Super Creepies**
Variety of liveries. Issued 1993.               Current price

**MT-320  Twelve-Pack Super Creepies**
Variety of liveries. Issued 1993.               Current price

**MT-340  Fact Cards**
Set of 24 fact cards with 2 new dinosaurs. Issued 1993.  Current price

**MT-350  Secret Single Pack**
Dinosaur. Issued 1993.                          Current price

**MT-360  Six-Pack Dinosaurs**
Second series. Variety of liveries. Issued 1993.   Current price

**MT-370  Twelve-Pack Dinosaurs**
Second series. Variety of liveries. Issued 1993.   Current price

**MT-860  Battle Card Game**
Enables the collector to play the In My Pocket Game. Each
battle card depicts a monster from the range with a score
for strength. You know the weakness, speed and
intelligence when you compete with your opponent to
win cards. The first person to win them all is the winner.
Set includes two monsters and a dice. Pack size 180 x 30
x 200mm.                                                        Current price

**MT-871 Screaming Monster Box**
The box, which growls when opened, is the ideal place
for your models; you can even buy an extra one and use
it as a scary lunch-box. Green with orange clips and
yellow and green eyes, although there is a version in pink
with blue eyes. Issued 1993.                                    Current price

MARKS & SPENCER

Although made by special arrangement with the Corgi Toy Company, the word
'Corgi' did not appear on the following toys. However, it was clearly possible
to see the superb quality of the toys as they were displayed on the counter. I
would like to thank the management of Marks & Spencer for their kindness in
helping with my research; my friend, Susan Pownal, the editor and tireless
worker for the Corgi Collector Club; and especially Adrienne Fuller, Marketing
Coordinator, Corgi Sales Office, Leicester.

**Small Set No. 8000  Formula One Racing**
With Elf Tyrrell and Hesketh F1. Black and white. Issued
1978. Deleted 1980.                                     £40    £20    £10

**Small Set No. 8001  Wings Flying Team**
With Lotus Elite and Tipsy Nipper on trailer. Issued 1978.
Deleted 1980.                                           £40    £20    £10

**Small Set No. 8002  Police Patrol**
With police Jaguar and Fiat X/9. White, yellow stripe and
blue livery. Issued 1978. Deleted 1980.                 £40    £20    £10

**Medium Set No. 8100  Racing Team**
With Land Rover, Fiat X/9 and JPS on trailer. Green and
white, yellow and black. Issued 1978. Deleted 1980.     £50    £20    £15

**Medium Set No. 8101  Wings Flying School**
Silver, black and red. Issued 1978. Deleted 1980.       £50    £20    £15

**Medium Set No. 8102  Breakdown**
Orange, white, yellow and white livery. Issued 1978.
Deleted 1980.                                           £50    £20    £15

**Large Set No. 8400  Racing**
Blue, white, green, white, black and gold. Issued 1978.
Deleted 1980.                                    £60    £25    £15

**Large Set No. 8401  Wings Flying Club**
Green, grey, silver, red, black, white, orange and white.
Issued 1978. Deleted 1980.                       £60    £25    £15

**Large Set No. 8402  Rescue Set**
Orange, white, yellow, white and blue. Issued 1978.
Deleted 1980.                                    £60    £25    £15

### F.W. WOOLWORTH

Made by Corgi. Here was a chance to give your collections a boost when Little
and Large were in the Woolworth's promotion of 1982. They signed some model
packets and boxes, which have a high value and should be valued by a toy expert.

**Gift Set No. 53; Specials Nos. 171, 172 and 173**
Released both as a gift set and as singles. Colourful design
and livery. Issued and deleted 1982.
Single                                           £20    £10    £5
Set                                              £75     —     —

**No. 1360  Batmobile**
Copper beige. Issued and deleted 1982. Limited edition.    £50    £20    £10

**No. 1361  Bond Aston Martin**
Copper beige. Issued and deleted 1982. Limited edition.    £50    £20    £10

**No. 1362  Bond Lotus Esprite**
Copper beige. Issued and deleted 1982. Limited edition.    £50    £20    £10

**No. 1363  Buck Rogers**
Yellow and black. Issued 1982. Deleted almost at once.
Limited edition. Good investment.                £50    £20    £10

**No. 1364  Space Shuttle**
Yellow and black. Issued and deleted 1982. Limited
edition.                                         £50    £20    £10

**No. 1384  Thunderbird**
Cream and orange and cream and black. Issued and
deleted 1982. Limited edition.                   £50    £20    £10

**No. 1396  Space Shuttles**
Black livery and decals. Issued and deleted 1982. Limited
edition.                                         £100   £40    £15

## CORGI

**No. 3  Missile and Land Rover Gift Set**
RAF blue with silver missile on trolley. Price 10/6. Issued
1958. Deleted 1963.                      £50     —     —

**No. 3  Modified Batman Gift Set**
Batmobile, Batboat and Batcopter. Price £2.50. Issued
1976. Deleted 1982.                  £40     £20     £10

**No. 4  Bloodhood Guided Missile Gift Set**
White and dark khaki. Price £1/1/-. Issued November
1958. Deleted 1961.                  £50     —     —

**No. 6  Rocket Age Gift Set**
Bristol Bloodhound on launcher, 350 Thunderbird GM on
trolley, 351 RAF Land Rover, 352 RAF staff car, 353 radar
scanner and 1106 Decca radar van. Price £4/10/-. Issued
August 1959. Deleted 1960.          £250     —     —

**No. 9  Corporal Missile Gift Set**
Red, white and khaki. Price £1/19/-. Issued November
1959. Deleted 1963.                  £20     £10     £5

**No. 40  The Avengers Gift Set**
John Steed's 3-litre Bentley, Emma Peel's Lotus Elan and
the figures of Steed and Peel. Price 16/11. Issued 1966.
Deleted 1969.                      £150     £50     £20

**No. 45  All-Winners Gift Set**
261 James Bond Aston Martin, 310 Chevrolet Sting Ray,
314 Ferrari Berlinetta, 324 Marcos Volvo, 325 Ford
Mustang competition. Price £1/16/-. Issued and deleted
1966.                          £100     —     —

**No. 117  Bloodhound Missile Trolley**
Olive-green. Price 9/6. Issued 1959. Deleted 1961.    £20     £10     £5

**No. 118  Thunderbird Guided Missile**
Silver-blue. Issued May 1958. Deleted 1963.        £30     £10     £5

**No. 155  UOP Shadow F1**
Black. Issued June 1972. Deleted 1976.           £40     £20     £10

**No. 156  Graham Hill's Special**
Lovely model for any collection, the Embassy Shadow
owned by the famous racing driver Graham Hill. White.
Issued June 1974. Deleted 1976.
Unsigned                         £50     £20     £10
Signed by Graham Hill               £2000     —     —

**No. C647  Buck Rogers Starfighter**
White with golden interior and with yellow and blue fins
and rocket tail. Two figures, one in silver and the other in
white. Price £1.95, although price varied. Issued 1982/84.
Deleted 1988.                                £50    £20    £10

**No. 1108  Bristol Bloodhound Guided Missile**
White missile on dark khaki ramp. Price 15/11. Issued
1958. Deleted 1961.                          £20    £10    £5

**No. 1109  Bristol Bloodhound Missile**
White and yellow missile on khaki trolley. Price 12/6.
Issued 1959. Deleted 1961.                   £20    £10    £5

**No. 1112  Corporal Guided Missile**
White missile on khaki launching ramp. Price 12/6.
Issued February 1959. Deleted 1961.          £20    £10    £5

**No. 1113  Corporal Erector Vehicle**
Red and white missile on khaki vehicle. Price 15/11.
Issued October 1959. Deleted 1962.           £20    £10    £5

**No. 1115  Bloodhound Guided Missile**
White livery. Price 10/6. Issued December 1959. Deleted
1961.                                        £20    £10    £5

**No. 1116  Bloodhound Missile Platform**
Dark. Price 5/-. Issued December 1959. Deleted 1961.   £5    £2    £1

**No. 1124  Corporal Guided Missile Ramp**
Khaki. Price 5/6. Issued 1960. Deleted 1961.   £5    £3    £2

**No. 1126  Batmobile**
Black with Batman logos. Figures of Batman and Robin.
Price 5/11. Issued October 1966. Deleted 1979.   £50    £10    £5

**No. 1128  James Bond Aston Martin**
This world-winner gave pride to the factory and the
workers who made it. In gold, it has three special features
including Bond and baddie. Winner of the 'Toy of the
Year Award for 1965'. Price 3/6. Issued 1965. Deleted
1969. Should you be lucky enough to have the autograph
of Sean Connery – correctly dated and witnessed – the
investment you have is something very special indeed.
Unsigned                                     £40    £20    £10
Signed                                       £2000   —     —

**No. 1130  Man from UNCLE Oldsmobile**
Lovely gift investment and a car of the future. Dark blue
or white (rare). Figures of Solo and Kuryakin. Price 5/11.
Issued 1966. Deleted 1969.

| | | | |
|---|---|---|---|
| Dark blue | £50 | £20 | £10 |
| White | £150 | £60 | £30 |

## FROM MAIL-ORDER CATALOGUES

**Argos, Thunderbirds Backpack**
Authentic livery and logos with adjustable padded
shoulder straps. Made in durable nylon. Price £4.99.
Issued 1992/3.                                      Current price

**Argos, Thunderbirds Talking Clock**
Ideal gift for the nursery, especially for the older children.
The famous countdown '5, 4, 3, 2, 1 Thunderbirds are Go'
keeps repeating after 10 minutes. Price £15.75. Issued
1992/93. First-class investment.            £30        —        —

**Bluebird Toys, Astro Shark**
In white, blue, red, orange and black, this five-in-one
spaceship has a 4-man control pack-deck, 4 machine-gun
stations, 2 drop bombs and 14 figures. Price £24.99. Issued
July 1993.                                          Current price

**Burlington, Captain Scarlet Mystery Game**
In brightly coloured box. Price £12.99. Issued July 1993.    Current price

**Burlington, Fist Fazer**
Silver, black and red with five robotic sounds, adjustable
strap and flashing light. Price £16.99. Issued 1992/93.    Current price

**Burlington, Gladiators Atmosphere Challenge**
Silver, blue, red and orange on large blackboard. Price
£19.99. Issued July 1993.                           Current price

**Burlington, Gladiators Danger Zone**
Series of toys in colourful array of liveries with wording
and motifs. Price £24.99. Issued July 1993.         Current price

**Burlington, Gladiators Super Duel**
Silver, black, red and blue. Shadow and contender fixed
onto swivelling podiums which turn when levels at the
base are moved from side to side. Four helmets and two
pugil sticks. Price £14.99. Issued 1993.            Current price

**Burlington, Jurassic Park Safari Vehicle**
Yellow, green, red line and orange wording. Price £19.99.
Issued 1993.                                        Current price

**Burlington, Sonic Man**
Silver, black and red. Eight electronic sounds with light,
three super interchangeable weapons and light-up eyes.
Price £11.99. Issued 1993. 9in. high.                    Current price

**Burlington, Ultimate Terminator**
Black, red or purple with motifs. Press the buttons on his
back and hear him talk. His eyes glow and flash, while a
machine gun makes realistic action sounds. Price £22.99.
Issued 1993.                                             Current price

**Grandstand Games, Thunderbirds**
Dodging the meteorites with four functional movements.
Price £9.50. Issued 1993.                                Current price

**Great Universal Stores, Thunderbirds Curtains**
Authentic colours and decals. Price £29.99. Issued
1992/93.                                                 Current price

**Great Universal Stores, Thunderbirds Outfit**
Authentic livery and decals. Price £14.99. Issued 1992/93.
Sound investment.                                        Current price

**Great Universal Stores, Thunderbirds Play Panels**
Red, blue and green and Thunderbirds decor, wording
and letters. Clip-together plaything for the nursery. Price
£29.99. Issued 1992/93.                                  Current price

**Great Universal Stores, Thunderbirds Playmat**
Price £15.99. Issued 1992/93. Very colourful collector's
piece to purchase and put away for a rainy day.          Current price

**Great Universal Stores, Thunderbirds Quilt Cover and
Pillowcase**
Price £25.99. Issued 1992/93. Another very colourful
collector's item to buy and store away as a future
investment.                                              Current price

**Great Universal Stores, Thunderbirds Valanced Sheet**
Authentic colours and decals. Price £21.99. Issued
1992/93.                                                 Current price

**Hasbro Toys, Almagator**
Purple and brown with green teeth and tusks. Issued
1991.                                                    £15    £8    £4

**Hasbro Toys, Bucky O'Hare**
This colourful character and his antics in space became
the favourite of millions of boys and girls from the age of
one. Red and yellow with green face. From the TV show.
Issued 1990/91.                                          £20    £10   £5

| | MB | MU | GC |
|---|---|---|---|

**Hasbro Toys, The Croaker Space Ship**
Blue and orange. Issued 1991/92.
£20 £10 £5

**Hasbro Toys, Dead-Eye Duck**
Orange and red. Character much loved by all Bucky
O'Hare fans. Issued 1991.
£20 £10 £5

**Hasbro Toys, Field Air Marshal Toad**
Brown spacesuit, green face, blue hat and arms. Issued
1991/92.
£20 £10 £5

**Ideal Toys, Electric Robotic Arm**
Blue, black and silver. Real hydraulic sounds, robotic
voice, commands, futuristic weapon sounds, barrel
which glows bright red, crushing metal sounds, controls
(including speed) and volume. Price £29.99. Issued
1992/93.
Current price

**Knickerbocker Toy Company, ET Collection**
In November 1982 this company announced that it had
secured the sole rights to make and distribute all soft toys
based on the characters from the film ET. First-class
investments.
ET only £100 £20 £10
Full collection £500 — —

**Knickerbocker Toy Company, ET Hand Puppets**
Issued 1982.
£50 — —

**Lott's Kindergarten, Flash Gordon Bricks**
One of the first coloured sets ever to be made, this is a
rare and hard to find investment of the adventures of
Flash Gordon's trip to Mars. Wood, although many
original brick playsets were made of stone. Instruction
story book. Price 6d for a box of 12. Issued 1936/37.
£200 £50 £10

**Peter Pan Playthings, Thunderbirds Rescue Game**
Blast off from Tracey Island and fulfil one of 16 missions.
Colourful game in colourful box. Price £7.99. Issued
almost exclusively to Argos in 1993. Good investment.
Current price

**Tomy Toys, Space Shuttle**
Red, white, blue and yellow. Two spacemen, one in a blue
suit and the other in green, with decals and the word
'Tomy'. Price £14.99. Issued 1992/93.
Current price

**Tracey Island Toys, Thunderbirds Playset**
Authentic livery and logos. Price £29.99. Issued 1992.93.
Current price

**Wesco, Stingray Cube Clock**
Battery operated. Moulded plastic with round corners
and edges. Dome shaped alarm shut-off button which
also activates the illuminating feature. Three sides of the
cube have stingray underwater scenes and six peel-off
transfers including Aqua Marina and Troy Tempest in
scuba gear. Price £8.99. Issued 1992/93. 4 x 4 x 4in.              £30         —         —

# JAPANESE
# SPACE TOYS

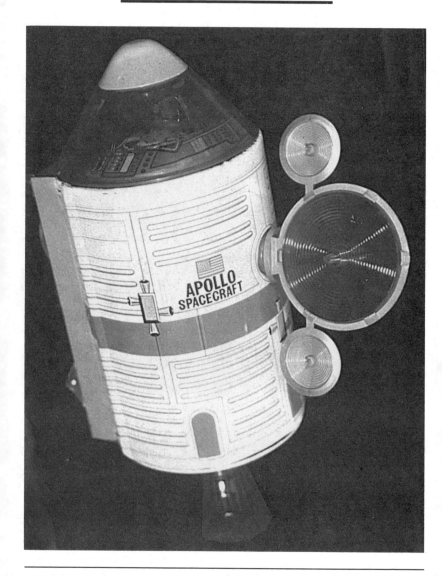

As explained in the Introduction, Japanese toys have become highly respectable and are among some of the most sought after in the world of collecting. Astronauts, rockets, satellites and tanks are among some of the toys listed below. There are still many others to be discovered and I would be glad to hear from any collector at any time.

| MODEL | MB | MU | GC |
|---|---|---|---|
| **Daiya Clockwork Astronaut**<br>Red with silver parts and backpack with straps of white. Blue helmet and black space gun. Issued 1960/65. Good investment. | £200 | £100 | £50 |
| **The PK Clockwork Astronaut**<br>All-silver, all-blue, all-red or all-green with the usual decor, dials and key mechanism either at the middle or at the lower region side. Issued 1955/65. All-green is worth treble, so watch out for this and any other rare colour. | £150 | £75 | £50 |
| **Wind-up Moon Astronaut**<br>Clever clockwork model which lifts up, fires and makes an Ack-Ack noise while in walking action. Mainly red with blue helmet and black weapon. Issued 1955/60. | £200 | £100 | £50 |
| **The Atom Boat**<br>Clever and unusual streamlined model with large tail and front driver. Issued 1965/70. | £100 | £50 | £25 |
| **The PK Bulldozer**<br>Battery operated in deep orange-red with light on nose, large front scraper and thick black rubber tracks. Blue engine seen on sides. Space robot driver in silver-blue or purple. Issued 1960/70. | £200 | £100 | £50 |
| **The Space Bus**<br>Red or maroon with space logo and the word 'Toys' on upper and lower side panels. Issued 1970. Rare. | £250 | £100 | £50 |
| **Capsule 5**<br>Battery operated in purple with red nose and red and green ovals around the front cone in silver-blue. Spaceman in dark orange suit, seen through clear plastic cockpit. The word 'Capsule' in large letters and the number '5' in red and white. Issued 1960. | £100 | £50 | £20 |

**Osaka Diecast Lead Rocket Car**
This special outer space model was made with the driver
cast in. On one side is 'Japan No. 1' and on the other
'Osaka UI'. The wheels are cast on one side of the axle and
spot soldered on the other. Inside the nose is a small
revolving propeller. If you blow down the air intake,
there are the sounds of a car and a factory buzzer. Issued
1969. Deleted 1986. 4½in. long x 2¼in. high x 1¼in. across
its beam.                                                          £350    £75    £45

**Osaka 301 Space Car**
Red or blue with white or yellow lines. Large white
wheels and black tyres. Battery operated driver with
flying sparks and sound. Price £9. Issued 1969. Deleted
1973.                                                             £200    £75    £40

**Space Car No. 3**
All-red with silver bumpers, screen and orange or lemon
lights. The number '3' on bonnet in black in blue circle.
Driver in blue space suit, all-friction drive with eyes that
open and close as car moves. Issued 1955.                         £50     £30    £10

**Space Car No. 8**
Red bonnet with white wide wheel panels and the
number '8' in black in white circle. Silver bumpers and
large wheels with thick black tyres. Space driver with
green suit and white square-type head with moving
lemon-orange eyes and silver antennae. Issued 1955.              £75     £40    £20

**King Jet Space Car**
Silver or blue with the words 'King Jet' and the number
'8' on sides. Driver and black wheels. Good
friction-sparking model. Issued 1955.                            £150    £75    £30

**Red Devil Space Car**
Red body with rocket gun in blue and white on bonnet.
Silver bumpers, silver wheels with thick black tyres, and
windscreen. Driver with silver suit and moving eyes,
friction action. Issued 1955.                                    £100    £50    £25

**The R-10 Space Patrol Car**
One of the best models ever made. Battery operated in
red and blue, and silver livery. With 'R-10' and the words
'Space Patrol' on sides, bonnet and nose. Silver wheels,
bumpers, grille and windscreen. Driver in silver space
suite and helmet. Issued 1965/70. Deleted 1973. Very
good investment.                                                  £300    £100   £50

**Space Car Solo Special**
Battery operated in red, silver and white with space robot
driver. Windscreen and white wheels with large black
tyres. Issued 1955.  £200  £100  £50

**The Apollo Space Craft**
Looks more like a large family-sized can of peaches on its
side. White. Scanner and antennae. Rear in red with
orange or red nose-tip. Battery operated. Issued 1965.  £50  £30  £15

**Apollo 1 Space Craft**
After three astronauts, Gus Grissom, Ed White and Roger
Chaffee, tragically died in the first attempt when a fire
destroyed their craft on the launching pad, several events
were to happen during 1967 to commemorate and pay
tribute to the brave men. US President John F. Kennedy
had given his blessing to any project when he said on 20
May 1961: 'I believe this nation should commit itself to
achieving the goal, before this decade is out, of landing a
man on the moon and return him safely to Earth.' Sadly
he was not to see the fulfilment of his ambition. This
Apollo 1 is in silver with US markings and decals. Price
£5. Issued 1967. Deleted 1970. Rare.  £250  £100  £50

**United Apollo Space Craft**
Battery operated in silver with large round fat body and
with orange and silver scanner and antennae. Red and
white or blue and yellow check design at rear. Issued
1965.  £50  £30  £15

**Combo 7 Space Craft**
All-action with spark and sound battery operated model.
Mainly silvery blue with white and orange rocket on rear
with clear plastic nose-tip. Silver wheels and black tyres.
Well armed with spacemen in silver suits and yellow
interior cockpit. White capsule top and the word
'Rendezvous' in thick red letters with white outline. The
numbers '7' in red and white on side and '8' on smaller
module in orange on black circle. Issued 1960.  £250  £100  £50

**The Commander Space Craft**
Well-designed, clever friction toy in a variety of colours
with driver commander and shark-fin tail. The word
'Commander' on sides. Issued 1960/70.  £50  £30  £10

**The Fire-bird No. 3 Space Craft**
Green with red and white line trim. The words 'Fire-Bird' in yellow and the number '3' in white on nose and in orange on the tail. Yellow, white, red and green rocket sides, driver in blue suit seen through clear plastic cover, and blue dot decals with black rings along top and lower. Issued 1960.

£200 £100 £50

**The Moon Detector Craft**
Battery operated in purple and silver with black rubber tracks beneath another can-shaped model, with silver arms, red and orange probes and view of spaceman driver. Blue and black dotted tractor. Issued 1960/70. Good investment.

£250 £100 £50

**Space Surveillance Craft X-07**
Yellow or orange streamlined nose with mainly blue streamlined design on base and sides. Driver with red suit seen through clear plastic driving area with flashing lights. Great movement action as model is switched on with sound. Issued 1965/70.

£100 £50 £25

**Planet Cruiser UN 751**
Battery operated in red, white, silver and black with white scanner and space driver with weapon. Issued 1960/65.

£100 £50 £20

**The Diesel Cultivator 12**
Large battery operated model in yellow or mustard with black or blue line design, grille and tracks. View of the working engine and '12', 'Cultivator' and 'Diesel' on top and rear sides in black and red. Robot driver in deep purple at controls. Issued 1960/70.

£250 £100 £50

**The Space Dog**
Red with mechanical action. Mouth opens and shuts while walking. Blue eyes. Black backstrap and silver tail with red nob tip. Issued 1965.

£200 £100 £50

**Planet Explorer**
Well-designed, battery operated model in mainly red or blue with driver in nose and scanner above with shark-fin tail. Issued 1965/70.

£100 £50 £25

**The Famous Mego Man Head**
Red, blue and orange, although colours can vary. The words 'Mego Man' in large black letters. Mechanical action. Issued 1955.

£100 £50 £25

**Martian Super Sensitive Patrol Jeep**
Battery operated in white or silver with orange or red wheels and thick black tyres. Open back type carries radar scanner, weapon, disc and rocket. Issued 1955. Very good investment.

| | MB | MU | GC |
|---|---|---|---|
| | £200 | £100 | £50 |

**The Count Down Machine**
Multicoloured design. Activated clock face. The pointers move and when they reach the correct number the rocket is released and fires upward. The words 'Count Down' in very large letters on base of model. Issued 1960/65.

| | £200 | £100 | £50 |
|---|---|---|---|

**The Builder Spaceman**
Clever mechanical battery operated model. Yellow with dark brown, chocolate or black head, arms and legs, yellow boots and matching helmet. Yellow barrow with black wheel. Red and silver control chest panel. Issued 1965.

| | £200 | £50 | £25 |
|---|---|---|---|

**Space Roundabout**
Colourful and clockwork roundabout of sorts with spacecraft and global capsules. The more the craft turns, the higher the ships go. Strong resemblance to real space models. Issued 1955/66.

| | £250 | £150 | £75 |
|---|---|---|---|

**The Saturn Missile SMR-7**
Friction operated in silver, red and black, although colours can vary. Chequered design and the words 'Saturn Missile, USA United States' on craft body. Issued 1960/70.

| | £50 | £30 | £10 |
|---|---|---|---|

**Space Module**
To commemorate Neil Armstrong's landing on the moon. Battery operated with seven automatic actions, one of which makes the module walk in an eccentric fashion, with antennae revolving, flashing lights and emitting a strange sound. The front hatch opens and shuts, and the spaceman model of Neil Armstrong emerges. Price £7. Issued 1968. Deleted 1970.

| | MB | MU | GC |
|---|---|---|---|
| Normal model | £350 | £100 | £50 |
| Signed by Neil Armstrong | £5000 | — | — |

**X17-878  Space Module**
Battery operated in silver with white or yellow-orange prop. Model moves with sound. Driver seen through clear cockpit. Issued 1965.

| | £50 | £25 | £10 |
|---|---|---|---|

**Space Module Z-206**
Prop action, battery operated in red, silver, gold and blue, although colours can vary. Chequered decor and painted figure on front. Issued 1965.

| | £50 | £30 | £10 |
|---|---|---|---|

**Osaka Space Monkey**
Very unusual battery operated model of a monkey with
two red suitcases. The figure walks, then at intervals
stops and somersaults over as suitcases stay supporting
the arms. Outer-space white suit with grey boots and
space helmet. Price £14. Issued 1968. Deleted 1976. £250 £100 £50

**Space Probe Model No. 2**
Battery operated with sound. Red body with large panel
and the words 'United States' in large white lettering.
Issued 1970. £50 £25 £10

**Space Game TR Lune One**
Multicoloured racing game which includes one small
space racer. When the clockwork is set in action, the fun
begins with great speed. World globe at larger end.
Issued 1965. £150 £100 £50

**Space Racing Game**
Battery operated in mainly blue with various designs on
the game board. Small silver rocket racer. Large round
piece with red, blue, green and pink flashing lights as
model is set into action. Issued 1965/70. £150 £75 £35

**The Robot Answergame**
Robot with brown head, yellow arms, grey body and
yellow base for purple lower parts, where number button
squares from 1 to 10 in 2 rows of 5. White. Issued 1960/70.
Unusual and rare. Good investment. £200 £100 £50

**Mystery Robot with Wide Top**
I have only ever seen one such model – in a collection I
valued. Silver, red, yellow and green. Issued 1965/70. £200 £100 £50

**Sparky Man Robot**
Battery operated in silver with red eyes, purple arms and
feet, and scanner on top of headgear. Issued 1965. £50 £24 £15

**Robotrac Special**
Battery operated in silver with dotted design vehicle with
tracks and with a large robot in driving seat. Very popular
in the early 1960s. £100 £50 £25

**The Space Rocker**
Purple with stardot decor and orange stars with brown
prop on top. Two gold robots on this clockwork model.
Rocks as the mouths open and close when movement
occurs. Issued 1960. £250 £100 £50

**Rocket TD 54**
Fuse-friction operated. Mainly white or silver, although
colours can vary. Chequered design. Issued 1960/70.  £50  £30  £10

**The Sonicon Rocket**
Battery operated, multicoloured model. Scanner and
driver seen quite clearly through the plastic cockpit
cover. Black wheels and spark action. Issued 1960.  £200  £75  25

**Rocket Firing Pad**
Silver with blue dots and silver wheel. Long purple,
yellow and silver firing section and silver and black
rocket with red fins and tip on nose. Issued 1960.  £50  £30  £15

**Rocket Launching Base Jupiter**
Battery operated in light mauve. Space controller in blue
suit in yellow control seat with TV screen in front and the
rockets ready to go. Yellow rockets with flame tips in
silver pad. Issued 1965/73.  £250  £150  £75

**Rocket Launching Base 2**
Red with white lines and black and blue insets. Black base
on silver tray. Many buttons, dials and other workable
parts with rocket, silver scanner and white tower or
column. Issued 1960/70.  £200  £100  £50

**Rocket Launching Base 3**
All-red base and lower battery operated section. White
or silver sides and red centre with large white dots and
thin circle line design. Several dials and switches with
blue scanner, white rocket with red base, and
white-edged tower with red light. Issued 1960/70.  £200  £100  £50

**Satellite X-10-7**
High quality battery operated model with distinctive
decor. Black and white triangle pattern of a weaving roll
of film and a plastic cup-disc on top. Light blue base and
dark wheels. Figure of spaceman can be seen through
window. 'Satellite X-10-7' in large orange letters and
numbers on lower side with yellow edging. Issued 1960.  £200  £75  £35

**Satellite X-11**
Battery operated in silver and blue. Red rimmed and clear
plastic dome cover revealing two space men in orange
suits and blue helmets with matching gloves. Further
spaceman in rear in dark suit, red boots and helmet.
Issued 1960.  £200  £100  £50

**Space Ride Clockwork**
This wonderful model has three space ships and as the
motion begins the ships fly out wider and wider. Clever
designs and logos. Issued 1960/70. Good investment.      £250    £100     £75

**Space Boy Ship X-15**
Battery operated, multicoloured model with plastic clear
cover revealing boy space pilot with 'X-15' on breast of
suit and the same decor on sides of ship. Issued 1960.    £150     £60     £20

**The Mars 107 Space Patrol Ship**
Large, attractive battery operated model with swivel
guns in red, yellow and silver with black or dark blue
barrels. Red front scanners in red and silver with
matching red top and bottom and wide silver centrepiece
with portholes and the words 'Space Patrol' in red. 'Mars
107' on lower red body in large white lettering. Issued
1960. Rare.                                               £350    £150     £75

**The Space Pioneer Ship**
Battery operated in silver. Silver wheels with black tyres.
Driver seen through clear plastic cockpit. The words
'Space Pioneer' in red or black on sides. Issued 1965.    £100     £50     £20

**The Money Bank Rocket Ship**
Mechanical special in silver, red, blue and white. Black
wheels, large at front and small at rear. Clearly worded
instructions on how to operate model. Arrow pinpoints
where the money should be inserted. Issued 1960. Rare.    £500    £200    £100

**Buddy L. Toys Space Shuttle**
Nice model from Japan. White and silver all-plastic
shuttle on a Mack truck with tinplate cab. Price £9,
although price varied. Issued 1976. Deleted 1983. Shuttle
150mm long.                                               £250    £100     £50

**Space Sightseeing Ship**
Very colourful battery operated model with side
windows and 6 passengers on each side of the craft with
the crew of 2 drivers in front. Issued 1965.             £200    £100     £50

**Moon Spaceship by PB**
Mainly silver or gold tinted with scanner detectors. Body
of the spaceship itself in silver-blue or purple with or
without spaceman driver. Some models have two
spacemen which can be moved in and out of seats. The
words 'Moon Space Ship' or 'Moon Patrol Space
Division' on sides. Good battery operated action with

sounds. Issued 1960/70. One of several produced at a period when many bargains were available but are now quite rare and always good investments. | £200 | £100 | £50

**The X-9 Special Spaceship**
Battery operated model with a larger than normal robot in deep blue with a purple tint seated on a craft in a well-designed livery. Plastic dome containing coloured balls which move about as model moves along with sound. Green front lights. Issued 1965/70. £150 £100 £50

**Spaceship X-1800**
Battery operated in white and red or blue, although colours can vary in this very popular type of craft. Scanner, shark-fin tail and ribbed front with driver. Issued 1960/70. £100 £50 £25

**LM Robotic Tractic Ship**
Battery operated in silver, red and black with thick black tracks of rubber. Issued 1965/69. 21in. long. £300 £100 £50

**The PB Shooting Range**
Clockwork game in a variety of colours and designs. Issued 1950/70, some of these models were among the first to be brought into other countries by illegal means or by tourists. Attractive collector's piece and hard to find. Boxes are rare. £250 £150 £75

**Moon Landing Sputnik**
Very interesting, well-designed and battery operated model. Colourful Earth globe in orange and blue with a long silver rod holding a sputnik in yellow, silver, blue, orange and brown. In the centre a silver rocket with fire-red nose. On the opposite end of moonbase is a blue, white and dark blue globe. Issued 1955. Special collector's item. £300 £100 £50

**Osaka Sputnik Special**
Multicoloured livery with a figure in space uniform wearing a spacesuit and helmet in silver. Price £14.75. Issued 1968. Deleted 1973. £100 £50 £20

**Space Tank**
Battery operated in purple and red with attractive rear orange and black squares and silver antennae with orange tips. Issued 1960. £50 £30 £15

**Space Tank**
Dark brown or fawn with white and blue edged front and red and white rear with silver tracks. Silver antennae with orange tips. Issued 1960. £50 £30 £15

**Space Tank Duo**
Battery operated with silver body and matching tracks
with two spacemen in front. Twin white rockets with
green tips. Sound and swivel action. Issued 1965.    £100    £50    £25

**Space Tank 6**
Clockwork operated in silver, red, white and black.
Caterpillar tracks with sound and spark-firing gun and
scanner. The words 'Space Tank' on sides. Issued 1960.    £100    £50    £20

**Area Radiation Tester**
Electrically operated in various colours with driver seen
through clear plastic cockpit and 'Planet Cruiser UN 751'
on sides. Issued 1965/70.    £100    £50    £25

# ROBOTS AND
# MOONWALKERS

Of all the wonderful and exciting things connected with space, I suppose that robots are well in line as to being favourite, especially the ones that do marvellous things at the flick of a switch. They are very collectable, and some of the early models can fetch a lot of money, like the rare Japanese Nomura 'Robbie Space Patrol' and the equally rare West German 'Dux Astroman'. The latter's empty box in itself is worth £200. Horikawa was a high quality company which turned out some of the best space toys. Colours vary, and I have done extra research on the following models listed, but there are still other variants to be found, so keep looking around and you could find a rare treasure.

| MODEL | MB | MU | GC |
|---|---|---|---|
| **AB3 Products, Bend-A-Bot Robot** Battery operated with controls on chest. Red or white body, arms and hands with yellow wrists and neck. Flexible head, moving arms, and flashing eyes and mouth. Price £7.50. Made in China exclusively for firm in Glasgow. 10in. high | | | |
| Red | £100 | £40 | £20 |
| White | £150 | £50 | £20 |
| **Alps Shoji, Television Spaceman** Multicoloured design. In original box. Aerial often missing with such models. Issued 1965/67. 10½in. high. | £250 | £50 | £20 |
| **Alps Shoji, Television Special 'R'** Battery operated. Lithographed tinplate television spaceman robot. In original box. 15in. high. Rare. | £500 | £200 | £75 |
| **Asakusa, Space Orbitor** Very unusual and rare toy to have, a space orbitor with astronaut. Silver, black and red. Issued 1965/68. 8in. long. | £100 | £40 | £20 |
| **Asakusa, Space Orbitor 2 with Astronaut** Special limited edition in red and gold, although other colours were made. Special gift box. Issued 1967. 10in. long. | £500 | £200 | £50 |
| **ATT Garibara, Robot on Motorcycle** A special release and rare collector's item in silver and red, with silver and blue robot, in special box. Issued 1970/75. 9½in. long. | £250 | £100 | £50 |
| **Brick Bradford, Robot** Silver, white, dark blue and brown with black boots and space weapon. Issued 1939, but rare because of Second World War. Found in US market, although still rare. | £2000 | £250 | £100 |

**Broham, Lunar Spaceman**
Silver plastic. Battery operated. Issued 1965/67. 11½in.
high.

£200 £50 £30

**Century 21 Toys, SWORD Moon Prospector**
Authentic livery. In special box. Issued 1965/67.

£50 £20 £10

**Century 21 Toys, SWORD Moon Ranger**
Authentic livery. In special box. Issued 1965/67.

£30 £10 £5

**Cragston, Astronaut**
Red with clear helmet, breathing mask, backpack and
cross belts in black and white decor. Figure has arms with
silver bands and orange gloves, holding black and purple
spacegun. The words 'Cragston Astronaut' in white
lettering. Large battery operated model. Issued 1955.

£500 £200 £50

**Cragston, Astronaut S-5**
Daiya battery operated, lithographed tinplate. Blue with
red and yellow detailing, oxygen tanks on back, with
plastic dome. Issued 1950s. 11in. high.

£200 £100 £50

**Cragston, Dalek with Easel Back**
Large and rare find in Dalek livery and logos. Issued
1965/69. 54in. high.

£200 £50 £30

**Cragston, Mr Atomic**
Blue with dark blue arms, yellow boots and red and black
striped waistband. Battery operated with sparking effect.
Issued 1955.

£100 £40 £20

**Cragston, LP Friction Explorer**
Became popular with all collectors of space. In special
box. Issued 1967/70.

£100 £50 £25

**Cragston, Robot**
Yonezawa battery operated. Lithographed tinplate
Cragston talking robot. Issued 1960s. 10in. high.

£200 £80 £40

**Cragston, Clockwork Robot**
Silver and light blue with crimped head, red ears and arm
joints, with thin red edge-line decor clock face dial and
red ring centre design. Very smart model. Issued 1955.

£100 £50 £25

**Cragston, Clockwork Robot**
Dark blue or purple with ribbed mouth, red ears and
orange line trim, with orange and black circle design at
front centre. Issued 1955.

£100 £50 £25

**Dux, Astroman**
Rare West German battery operated, four button remote
control, green plastic body with flashing lights, walking,
bowing, arms raising and opening movements. Original
box with instructions and aerial in original envelope.
Issued early 1960s.

£900 £350 £100

**Excelo, Space King Robot**
Battery operated lithograph. Blue and silver livery,
although colours may vary. Issued 1968. 11½in. high.

£150 £75 £35

**T. Hong Kong Toys, Moon Ranger**
Colourful and well-designed model. Silver or red in
special box. Issued 1970s.

£50 £20 £10

**Hong Kong Toys, Special Grey Plastic Robot**
Hong Kong battery operated, grey plastic robot with
opening chest to reveal guns.

£100 £75 £25

**Horikawa, Attacking Martian Robot**
Green and gold. Much sought after Japanese variant.
Booklet and special box. 15in. tall.

£350 £150 £75

**Horikawa, Attacking Martian Robot**
Battery operated, lithographed tinplate model. Brown
with red feet, opening chest with guns. In original box
with Japanese script. Issued 1960s. 14½in. high.

£250 £100 £40

**Horikawa, Attacking Martian Robot**
Battery operated. Colourful livery which can vary. Rare
in original box. Issued 1964/67. 11in. high.

£200 £50 £30

**Horikawa, Attacking Martian Robot**
Brown. Martian body with space astronaut's head. Issued
1965/67. 11½in. high. Good investment.

£200 £50 £30

**Horikawa, Dino Robot**
Battery operated, lithographed tinplate Dino robot with
splitting head to reveal dinosaur head. Blue. Issued late
1960s. 12in. high.

£500 £200 £75

**Horikawa, Dino Robot**
Battery operated, lithographed tinplate and plastic Dino
robot with walking action and flashing Dino head. Red
and black. In original box. Late 1960s.

£750 £250 £100

**Horikawa, Dino Robot**
Large robot with splitting head to reveal dinosaur head.
Head in green and flame-red, body in black or brown.
Issued 1968. 12in. high.

£300 £100 £50

| ROBOTS AND MOONWALKERS | MB | MU | GC |
|---|---|---|---|
| **Horikawa, Engine Robot**<br>Tinplate. Blue, silver, red and black with black and gold lines. Issued 1965/68. 9in. high. | £100 | £50 | £30 |
| **Horikawa, Super Face Robot**<br>Battery operated, lithographed tinplate super space giant robot. Black and red with rotating body, opening doors in chest to reveal gun, in original box. Issued 1960s. 16in. high. | £350 | £150 | £75 |
| **Horikawa, Plastic Flashing Robot**<br>Silver, red and blue, although colours can vary. Issued 1965/68. 9in. high. | £100 | £40 | £20 |
| **Horikawa, Gear Robot**<br>Red, green and blue. Issued 1965/70. 9¼in. high. | £150 | £75 | £35 |
| **Horikawa, Gear Robot**<br>Silver, blue and black. Issued 1965/68. 11in. high. | £100 | £40 | £20 |
| **Horikawa, Clockwork Plastic Gear Robot**<br>Green, mottled gold and black. Issued 1967/69. 8in. high. | £100 | £40 | £20 |
| **Horikawa, Giant Robot**<br>Battery operated, lithographed tinplate and plastic model. Black and gold. Issued 1960s. 16in. high. | £300 | £100 | £40 |
| **Horikawa, Giant Tinplate/Plastic Robot**<br>Another special find. Black or brown. Issued 1965/70. 15in. high, although some are 16–18in. high. | £200 | £100 | £50 |
| **Horikawa, 'H' Golden Line Robot**<br>Battery operated tinplate Attacking Martian. Gold. In original box with inserts. Issued 1960s. | £250 | £100 | £40 |
| **Horikawa, Piston Robot**<br>Liveries vary. Gold lines. Issued 1965/68. 10in. high. | £100 | £40 | £20 |
| **Horikawa, Roto Robot**<br>Yellow, green and black. Issued 1965/68. 9in. high. | £100 | £40 | £20 |
| **Horikawa, Roto-Matic Robot**<br>Battery operated Rotate-O-Matic Super Astronaut robot. In original box. 12in. high. | £250 | £100 | £50 |
| **Horikawa, Japanese Novelty Toy Robot**<br>Metallic blue, battery operated, lithographed tinplate and plastic radar robot. 12¾in. high. | £150 | £75 | £35 |

| | MB | MU | GC |
| --- | --- | --- | --- |
| **Horikawa, Vision Robot**<br>Battery operated, painted and lithographed tinplate model. Green and black. Issued 1960s. 11¼in. high. | £300 | £100 | £40 |
| **Horikawa, Vision Robot**<br>Battery operated, painted and lithographed tinplate model in limited production. Colours vary. Issued 1965/68. 11¼in. high. | £250 | £100 | £50 |
| **Horikawa, Space Scout**<br>Orange and grey with space markings. Issued 1965/70. 9¼–10in. high. | £250 | £100 | £50 |
| **Horikawa, Super Space Capsule**<br>Tinplate. Brown, silver and gold, although colours can vary. Issued 1967/69. 12½in. long. | £200 | £100 | £50 |
| **Kanto, Father Christmas Robot**<br>Battery operated. Red, white and silver. Limited edition for Christmas 1986. 8in. high. | £500 | £200 | £100 |
| **Kanto, Rear-Winding Friction Robot**<br>Unusual model to own. Lithographed tinplate robot with eccentric movements, in blue-silver and black, although colours can vary. Issued 1965/69. 10in. | £250 | £100 | £50 |
| **Kanto, Tinplate and Rubber Robot**<br>Clockwork lithographed tinplate, marked 'TT'. Black and red. Issued 1970/75. Rare. | £200 | £100 | £40 |
| **Kosuge, Black Robot**<br>Clockwork lithographed tinplate high-wheel robot. Issued 1960s. 11in. high. | £200 | £80 | £40 |
| **Kosuge, Black Clockwork High-Wheel Robot**<br>Lithographed tinplate. Silver lines. Rare livery with all working parts, a find for any collector. Issued 1965/68. 11in. high. | £350 | £100 | £50 |
| **Kosuge, Blue Robot**<br>Battery operated (by remote control), lithographed tinplate high-wheel robot. 11in. high. | £200 | £90 | £40 |
| **Kosuge, Blue Robot**<br>Battery controlled and operated, lithographed tinplate high-wheel robot. Blue with gold and silver stars. Issued 1965/68. 11in. high. | £250 | £100 | £50 |

**Kosuge, Blue and Silver Robot**
Battery operated (by remote control), painted and
lithographed high-wheel robot. Metallic blue. In original
box. Issued 1960s. 9½in. high.     £300    £100    £50

**Kosuge, Red and White Robot**
Tinplate and plastic clockwork Action Planet robot. In
original box. Issued 1960s. 8½in. high.     £250    £100    £40

**Larami, Super Soakerman No. 9896**
Green and yellow with red top, goggles and spacegun.
'Supersoaking Power in the Palm of His Hand.'
Articulated neck, arms and legs. Internal water reservoir.
Powerful pump-shot action. In colourful box. Issued 1991
in limited number. Made in China for Larami Corp.,
Philadelphia, USA. 12in. high.     £75    £40    £20

**P.B. Linemar, The Smoking Robot**
Fine working model with light-up eyes, 'stop and go'
action, lighting pistons and the very ingenious smoking
action. Mainly in purple with silver tint, although colours
can vary. Attractive box, which is rare. Issued 1960/70.    £250    £100    £50

**Marx, Dalek**
Silver with Dalek markings and livery. Issued 1965/68.
6½in. high.     £200    £100    £50

**Marx, Two Stingray Model Robots**
In double gift presentation box, although the models can
be found separately. Battery operated (by remote
control). Issued 1965/68. Rare.
Single model     £100    £50    £30
Both models     £500    —    —

**Masudaya, NASA Astro Captain Model**
Rare clockwork robot with face peering through clear
plastic helmet in silver and red or silver and blue. Issued
1960.     £100    £50    £25

**Masudaya, Mighty 8 Robot**
Battery operated, painted tinplate. Issued 1960s. 11½in.
high.     £200    £60    £30

**Masudaya, Target Robot**
Battery operated, lithographed tinplate. Issued 1960/65.
15in. high.     £800    £399    £100

**Masudaya, Target Robot**
This is the larger robot in black, red and silver. Battery
operated, lithographed tinplate. Issued 1967. 18in. high.    £750    £200    £50

**Mr Backpack Moonwalker**
Silver, black and gold with clockwork action, which goes
from side to side. Issued 1970.

|  | £100 | £50 | £20 |

**Noguchi, Space Robot**
Lithographed tinplate space man. Green and orange
mighty robot with sparking chest compartment. 5½in.

|  | £250 | £100 | £40 |

**Nomura, Blue Rosko Astronaut**
Battery operated, lithographed tinplate astronaut,
similar body-pressing to Nomura Robbie, with perspex
parabolic dome buzzer and light in helmet, rubber arms,
walkie-talkie with light, coil antennae, backpack and
walking action. Issued late 1950s. 13in. high.

|  | £500 | £150 | £75 |

**Nomura, Earthman**
Battery operated tinplate, with red remote-control box,
sounding and blinking gun, and retractable aerial.
Yellow, silver and red. In original box. Made in the 1950s.
9in. high.

|  | £750 | £250 | £50 |

**Nomura, Robot**
Battery operated, mechanised tinplate robot. Black and
red with clear plastic parabolic dome and rubber hands,
with circular handles on battery-cover latches. Issued
1950s. 13in. high.

|  | £450 | £150 | £65 |

**Nomura, ASKA Tom-Toy Institute Mechanised Robot**
Battery operated, based on Robbie the Robot. In special
box. Red, silver and black, although other colours known.
18in. high.

|  | £250 | £150 | £50 |

**Nomura, Robbie the Robot**
Mechanised, painted tinplate and plastic. Gold with
black lines. In special box. Issued 1955/60. 13¼in. high.
One of the most sought after models ever made for the
world market.

|  | £1000 | £500 | £100 |

**Nomura, Robbie Space Patrol**
Battery operated, lithographed tinplate mechanised
robot with mystery action, clear plastic dome, light dishes
and rubber hands. Blue, gold and red. Issued late 1950s.
12½in. long. Rare.

|  | £5000 | £1000 | £250 |

**Nomura, Robotank Z Robot**
Battery operated, lithographed tinplate robot. Green.
Issued 1960s.

|  | £300 | £100 | £50 |

**Nomura, Robotank Z Robot**
Battery operated, lithographed tinplate. Bronze and black. Issued 1965/68. Rare find in mint and boxed condition.

| | £350 | £100 | £50 |
|---|---|---|---|

**Nomura, Zoomer the Robot**
Battery operated, painted and lithographed tinplate model. Red and gold. Issued 1955/60. 9in. high.

| | £300 | £100 | £50 |
|---|---|---|---|

**Paget Bros, Saturn TV Robot**
Battery operated walking robot with light-up eyes and shooting missiles. Large figure with black or dark brown body with red eyes, claws and feet. Activating screen on chest with various scenes of outer space. In colourful box. Price £10. Made in China for Paget Bros, Wakefield, Yorks.

| | £100 | £50 | £20 |
|---|---|---|---|

**PK Toys, The Z Man**
Clockwork man seated on top of rocket, well coloured in white, red, silver, black and gold, with tints of blue, although colours can vary. With Z on tail and Z Man in centre. Issued 1960/65.

| | £150 | £75 | £40 |
|---|---|---|---|

**Popy, Space Robot**
Blue and white with slight silver tint all through the model design, although other colours known. Issued 1970. 11in. high.

| | £250 | £100 | £50 |
|---|---|---|---|

**Popy, Robot No. LS21  Space Robot**
Japanese. Black with silver boots, leggings and helmet. Red armlets, silver gloves.

| | £100 | £40 | £20 |
|---|---|---|---|

**Popy, Robot No. LS22  Space Robot**
Japanese. Red with boots, leggings and armlets. Silver gloves and blue helmet top.

| | £100 | £40 | £20 |
|---|---|---|---|

**Popy, Robot No. LS23  Space Robot**
Black body, boots and leggings with red stripe, red armtops and gloves. Silver armlets, gold horns, and red, black and silver shield. Price £5.

| | £200 | £75 | £35 |
|---|---|---|---|

**Popy, Robot No. LS25  Space Robot**
Yellow body with black leggings and boots with red stripes. Black and silver armlets and red gloves. Price £5.

| | £200 | £100 | £40 |
|---|---|---|---|

**Rosuge, Plastic Walking Robot**
Tinplate. Green, black, silver and gold with black lines. Model in mint and working condition is rare. Issued 1967. 11in. high.

| | £250 | £100 | £50 |
|---|---|---|---|

| | MB | MU | GC |
|---|---|---|---|
| **RST, Horned Robot**<br>Lithographed tinplate and plastic model. Blue and yellow. Issued 1970/75. 8½in. high. | £100 | £40 | £20 |
| **Sconoscuito, Clockwork Sparking Robot**<br>Rare and colourful early robot with markings and space details. Silver st?¬s. Issued 1954/55. | £350 | £100 | £40 |
| **Strenco, Clockwork Robot**<br>Tinplate ST-1. Red, silver and black. Issued 1955/57. 7¼in. high. Rare. | £200 | £50 | £30 |
| **Strenco, Clockwork Painted Robot**<br>Super model in colourful design. Tinplate ST-1. Issued 1967/69. 7¼in. high. | £200 | £50 | £30 |
| **Taiyo, Green Robot**<br>Battery operated, painted tinplate and plastic Wheel-A-Gear robot. Issued 1960s. 14½in. high. | £250 | £100 | £50 |
| **Tomy, Robot Man R7**<br>Battery operated. Purple with red ears, boots and claw hands, with sparking and red centre dial with the R7 markings in orange. Issued 1955. Deleted 1960. | £100 | £40 | £20 |
| **Taiyo, Target Robot**<br>Battery operated, lithographed tinplate robot. Silver and black with red target on chest causing evasive action, with illuminating mouth and swinging arms. Issued 1950s. 15in. high. | £1000 | £350 | £100 |
| **Tomy, KI-KU-ZO Voice Red-RD Recognition Robot**<br>Battery operated, talking robot, with voice recognition system, with instruction booklet, and a silver and black flat printed car. Issued 1970/74. Rare. | £200 | £50 | £30 |
| **Tomy, Silver Spaceman**<br>Another for the specials in silver with red or blue lines and dotted effect around base on which the figure stands. Issued 1973. 7in. high. | £200 | £100 | £50 |
| **Tomy, Torchy the Battery Boy (in Space Suit)**<br>Chinese. In authentic livery and markings as taken from the TV series. Price £3.00. Issued in 1965. Deleted almost at once. | £100 | £50 | £20 |
| **Yanoman, S17 Space Scout**<br>Red, blue, black and orange, although colours can vary. Issued 1967. 9¾in. long. | £200 | £50 | £30 |

| | MB | MU | GC |
|---|---|---|---|
| **Yonezawa, First Version of Lost in Space Robot**<br>Black and silver. Lost in Space logo and markings. Issued 1965/67. 9in. high. | £250 | £100 | £50 |
| **Yonezawa, Second Version of Lost in Space Robot**<br>Painted livery and colours which can vary. Issued 1967/69. 10in. high. | £100 | £50 | £20 |
| **Yonezawa, Delta 55 Lunar Explorer Robot**<br>Battery operated. Colourful livery, although colours can vary. In special box. Issued 1968/72. | £50 | £20 | £10 |
| **Yonezawa, Space Explorer Robot**<br>Battery operated tinplate model, transforming from a TV set into a walking robot with flashing eyes and with astronaut pilot in chest panel. Red. Issued 1960s. 18in. high. | £500 | £200 | £75 |
| **Yonezawa, Sparking Robot**<br>Multicoloured livery and manufacturer's markings. Issued 1968. 7in. high. | £150 | £75 | £35 |
| **Yonezawa, Talking Robot**<br>Battery operated, friction drive tinplate robot. Designs can vary. In original box. Issued 1960s. | £750 | £250 | £100 |
| **Yonezawa, Talking Robot**<br>Battery-operated, friction drive tinplate robot. Red and silver. In original box. Issued 1960s. | £500 | £200 | £75 |
| **Yonezawa, Clockwork Walking Robot**<br>Lithographed. Black, blue, gold and orange. Company logo markings. Issued 1966. 10in. high. | £200 | £50 | £20 |
| **Yoshiya, Orbit Man and Orbit Explorer**<br>With airborne satellite (9in. long). Rare instruction leaflet. | £250 | £100 | £40 |
| **Yoshiya, Chief Robot Man**<br>Battery operated, painted tinplate and plastic robot. Silver and blue. Instruction leaflet. Issued 1960s. 12in. high. | £250 | £100 | £40 |
| **Yoshiya, Chief Robot Man**<br>Another multicoloured, painted tinplate and plastic collector's gem. Issued 1965/68. 12 or 16in. high. | £250 | £100 | £50 |
| **Yoshiya, Clockwork Robot**<br>Lithographed and sparking. Silver.Issued 1967. 7in. high. | £100 | £40 | £20 |

**Yoshiya, Clockwork Robot**
Lithographed and sparking. Red and Blue. Other colours
are rare. Issued 1967/69. 10in.

| | MB | MU | GC |
|---|---|---|---|
| Yoshiya, Clockwork Robot | £100 | £50 | £20 |

**Yoshiya, Planet Robot**
Clockwork, painted tinplate robot. Metallic grey with red
hands and feet, and chromed face grille and chest area.
Issued 1960/65. 9in. high. Rare.  £450  £200  £75

**Yoshiya, Spaceman Robot**
Clockwork lithographed tinplate. Silver and black. 8½in.
high.  £200  £75  £40

**Yoshiya, Clockwork Spaceman Robot**
Lithographed tinplate with splendid clockwork
movement. Black and red with decals and markings.
Issued 1965/70. 8½in. high.  £200  £100  £50

# GUNS

This book would not be complete without the inclusion of the weapons which made the Flash Gordons, the Buck Rogers and the Star Trek or planet commanders safe from the creatures of outer space. I have chosen the best of what has been made – 'created' would be an even better word – in the vast number of space guns, laser guns and other spark-flashing weapons which are now highly collectable and mean so much to the millions of enthusiasts in the elaborate world of firearms that originated in the 1930s. I would like to draw attention not only to an item such as the American Han Solo Star Wars Laser Pistol, but to the Hong Kong Invincible Space Weapon. The latter represented the start of a more superior range of weapons from such places as China, Hong Kong and Taiwan; these weapons were rather looked down on at first but later became highly respected and much sought after. The research for this section was a delightful period for me, and I would like to thank the many people who sent me information. In particular I must thank 'R' for her kindness, friendship and confidence, not to mention her gracious company, especially in the year when we travelled 29,000 miles.

| MODEL | MB | MU | GC |
|---|---|---|---|

**The Astroray Gun (Shudo, Japan)**
Silver with spark seen through square-type window at front of topside of weapon. The words 'Shudo Astroray Gun' in black and white. Other designs in orange, red, lemon and black. Orange and yellow and black line design on main handle grip. Silver trigger. Issued 1977/79.

£100 £40 £10

**The Astroray Gun (Shudo, Japan)**
Purple, red, white, yellow and white with black trigger. The words 'Astroray Gun' in orange-red with yellow at top. Issued 1987/89.

£50 £20 £10

**The Atom Ray Gun (Hiller, USA)**
All red with silver barrel and screw-type end. Looks like a tiny half-Zeppelin. Issued 1935/37.

£250 £75 £30

**Atomic Flash Gun (J. Chein)**
Blue or green with yellow or orange trim and red lettering, with flashes. Issued 1956.

£50 £20 £10

**Batman Bat Ray Projection Pistol (Remco, Newark, New Jersey, USA)**
Black with authentic Batman livery and decals. Issued 1977.

£50 £30 £10

**Batman Pistol and Rifle Set (Hong Kong)**
This set was on sale in the Birmingham expo-type fair 1978. Only a few available in black, red and yellow authentic Batman logo-design. Rare and probably very limited issue. Excellent investment.

£1000 £200 £50

**Black Special Space Gun (Japan)**
One of the finest space guns ever made. Beautiful black colour and white dotted design and lines with sparks striking the red glass bead and front design. White handle with black lines and yellow edge trim. Words in orange and pink decor. Issued 1956.

£100 £50 £20

**Blue and Black Ray Gun (China)**
Blue, black and silver, bulb shape. More common all-blue model was also made and prices slightly lower. Issued 1987/89.

£30 £10 £5

**The Blue Laser Gun (Azrak Hamway International, Hong Kong)**
Blue with yellow trigger and clear, short part barrel with flash. Bulb-type body popular at the time. Issued 1978.

£50 £20 £5

**The Blue Sparker (Hong Kong)**
Almost clear plastic model with purple or blue tint, red
stopper end and silver trigger. Issued 1980, although
earlier models have been reported.

**The Bub-L-Rocket Gun (Kenner, Cincinnati, Ohio, USA)**
Yellow and blue, although colours vary. Matching rocket
ship was also made as a set, which is rare. Issued 1968/69.

| | MB | MU | GC |
|---|---|---|---|
| The Blue Sparker | £50 | £20 | £5 |
| Gun | £100 | £50 | £20 |
| Rocket ship | £50 | £20 | £10 |
| Set | £250 | — | — |

**Buck Rogers Disintegrator 1936**
Red, blue and orange or gold with fat bulb-shaped body,
with spark. Rare. — £750 £200 £50

**The Buck Rogers Rocket Pistol (Daisy, Plymouth, USA)**
Silver-grey. Issued 1936. — £750 £200 £50

**Buck Rogers 25th Century Rocket Pistol (Daisy, Plymouth, USA)**
Blue and silver or black and silver. Special box with a
picture of Buck Rogers on the cover. Price US50c. Issued
1936. — £750 £200 £50

**Buck Rogers XZ-38 (Daisy, Plymouth, USA)**
Metallic blue. One of the classic disintegrators with a flint
in the top which made a spark that you could see through
the 'electronic' compression viewplate. Issued 1936. — £750 £200 £50

**Buck Rogers XZ-44 Water Pistol (Daisy, Plymouth, USA)**
Well-made liquid helium water gun. Yellow or gold with
red or orange design on the handle and front barrel. Rarer
copper finished item worth treble. Issued 1936. — £500 £200 £50

**Captain Video Secret Ray Gun (Candy Bar)**
Red. In special box with battery, bulb and Luma-Glow
card for writing secret messages. Space map showing
relative size and distance from the sun at centre of map.
Issued 1956/58. — £100 £40 £10

**The Dynamic Space Gun (Hong Kong)**
This all-silver, bulb and diamond shaped gun is unique
in the toy world. Clever and attractive with sounds and
flashing lights. Issued 1987 (although earlier types have
been reported). — £250 £200 £50

| GUNS | MB | MU | GC |
|---|---|---|---|

**Expo-Gun (Barrline Products, Hong Kong)**
Black and lemon with black trigger, pictures and logos.
This long sparking weapon was larger than normal.
Issued 1978.

| | £100 | £50 | £20 |
|---|---|---|---|

**The Flash Gordon Blaster**
Red and silver with painted picture of Flash Gordon on
the front. Issued 1940.

| | £250 | £100 | £50 |
|---|---|---|---|

**Flash Gordon Click Ray Pistol (King Features Corp.,
USA)**
Red plastic with yellow trigger. Colourful box with
pictures of Flash Gordon firing at enemy. Issued 1955.

| | £250 | £50 | £10 |
|---|---|---|---|

**Flash Gordon Gun and Holster Set (Lone Star No. 1210)**
White, blue and red. Price 65p. Issued 1982. Deleted 1986.

| | £125 | £40 | £15 |
|---|---|---|---|

**Flash Gordon Pistol with Flash (Nasta, Hong Kong)**
Mainly silver with red trigger. Unusual shape with large
red and black tip at end of barrel. The words 'Flash
Gordon' on red oblong in gold letters, and his picture on
top sides. Issued 1972.

| | £100 | £40 | £10 |
|---|---|---|---|

**Flash Gordon Space Gun (Hong Kong)**
Silver with black muzzle gun end and black trigger. The
words 'Flash Gordon' in red and yellow on blue
background, and his picture on handle. Issued 1979.

| | £40 | £20 | £5 |
|---|---|---|---|

**Flash Gordon Space Pistol (Lone Star No. 1080)**
Flash Gordon has been acclaimed as the greatest
American hero of all time by many American and other
world press reports. Portrayed by Larry Buster Crabbe
along with his co-star, Dale Arden, in full-length movies
and in a BBC/TV series. This gun has a strong diecast
metal cap repeating mechanism for 100-shot roll caps. A
special feature is the transparent blister surrounding the
cap chamber area which is illuminated by the flash from
the exploding caps. Packed in attractive box. Price 50p,
although price varied. Issued 1982. Deleted 1986.

| | £100 | £30 | £15 |
|---|---|---|---|

**Flash Gordon Space Rifle**
This very rare and much sought after large space weapon
came in red, blue, black and silver, although colours vary.
Pictures of Larry Buster Crabbe and Dale Arden on front
of box, which alone is worth a high price. Issued 1938.
Limited edition.

| | £500 | £200 | £50 |
|---|---|---|---|

**Flash X-1 (Shudo Toys, Japan)**
Highly designed model mainly in red with orange and
blue stripes on handle. Yellow, blue and green design and
silver trigger. Issued 1966.

| | £50 | £20 | £10 |
|---|---|---|---|

| GUNS | MB | MU | GC |
|---|---|---|---|
| **The Han Solo Star Wars Laser Pistol (Kenner, Cincinnati, Ohio, USA)** Mainly black or very dark grey with the Star Wars decals and wording. Issued 1978. | £100 | £40 | £20 |
| **The Hubley Atomic Disintegrator** One of the best ray guns of its period. Metallic blue with red handle and cap repeater. Issued 1940s. | £200 | £50 | £20 |
| **The Indiscriminate Attacker Pistol** No marks on this model. Silver and gold with red and black ring design. Matching holster sold separately. Issued 1968. | £250 | £100 | £50 |
| **The Invincible Space Weapon (Hong Kong)** Issued 1970. Very rare. | £2000 | £200 | £100 |
| **James Bond Presentation Set (Lone Star No. 1211)** 007 pistol, silencer, shoulder holster, binoculars, hand grenade and decoder. Price £4.50. Issued 1982/83. Deleted 1986. | £75 | — | — |
| **The Laser Gun (Azrak Hamway International, Hong Kong)** Silver and red with blue solid trigger. Issued 1985/88. | £40 | £20 | £5 |
| **Moonraker Pistol 8-Shot and Flippy (Lone Star No. 1207)** Modelled on the pistol used by James Bond in the film Moonraker. Strong diecast metal body and mechanism. Price 50p. Issued 1982. Deleted 1986. | £100 | £40 | £20 |
| **Moonraker Pistol (100 Shot) (Lone Star No. 1208)** White and red finish. Price 50p. Issued 1982. Deleted 1986. | £60 | £15 | £10 |
| **The Multisound Ray Gun (Japan)** This unusual gun with no markings was definitely made by a Japanese firm. Silver-grey. Red and green ball end flasher with sounds. Thick black trigger. The unusual shape gives this model its high price. Box is important. Issued 1988/89. | £200 | £50 | £20 |
| **Multisound Space Gun (Hong Kong)** No markings, but I found it in 1987 among a whole selection of imported models from Hong Kong. Silver with orange tip horn shape, and red and blue on top. Unusual shape, streamlined for its time. | £75 | £35 | £15 |

**The Pencil Sharpener Ray Gun**
Unusual model. Red with tubular shapes on handle, with black inset and trade mark on barrel. Issued 1966/68. Rare. — £200 / £50 / £20

**Pressure Dragster Space Gun (China)**
Silver and red with silver trigger. The word 'Dragster' and picture of space module and rocket ship. Weapon has flash and sound. Issued 1989. — £100 / £50 / £20

**Radar Gun (Japan)**
All-silver with yellow design figure of spaceman uppermost and part of the trigger system in yellow or gold. Issued 1964. — £50 / £20 / £10

**The Ray Rifle (China)**
Long rifle-type weapon with mansize wooden livery handle, black centre-piece and white, red and orange tipped thinning-out barrel, with sound. Issued 1986/89. Good investment. — £100 / £40 / £20

**Razor Ray Gun (HY Manufacturing, Hong Kong)**
Gold, red and silver with large screw-type shape in ring form around body, silver trigger and green side button, with maker's markings. Issued 1971. — £50 / £20 / £5

**The Red/Green Sparking Gun (WF, Hong Kong)**
All-red or all-green with black trigger. Issued 1985/89. — £20 / £10 / £5

**The Reflectivity Pistol (Hong Kong)**
This very unusual weapon in bright yellow and silver has a mirror-type reflector on top of the barrel in conjunction with the trigger with black outline and orange stars. Issued 1974/75. Rare. — £500 / £100 / £40

**The Researcher Space Weapon (Homeline Models, Hong Kong or Japan)**
No markings apart from the words 'Home Product' on silver and black space rifle with pictures of clouds, rockets and two spacemen in a buggy. Issued 1976. Rare. — £1000 / £200 / £50

**Shining Light Gun (Vanity Fair Industries, USA)**
Silver with red trigger and screwline-type barrel end. Huge bulb-like shape with light flashing system. Issued 1978. — £50 / £20 / £10

**The Silver Midget Space Gun (Blue Q Enterprises)**
Silver. Issued 1983/88. Rare — £100 / £40 / £10

**The Silver Ray Gun (Korea)**
Very much shorter weapon than the normal product.
All-silver with red front-top and black button-type screw.
The first of many from Korea in this class. Issued
mid-1980s.

| | MB | MU | GC |
| --- | --- | --- | --- |
| | £50 | £20 | £10 |

**The Smoke Ring Space Gun (Nu-Age Product
Creations, Beverly Hills, California, USA)**
Grey or light metallic blue with red circle on nose and in
the centre. Supplied with special smoke matches shaped
like rockets to place in the chamber of the gun. An air
bladder blew perfect smoke rings from the barrel. Issued
late 1940s and early 1950s. £150 £75 £35

**Space Control Gun (Japan)**
Mainly light green with orange, red, lemon and silver
design and wording. Picture of spaceman and rocket
design on handle. Issued early 1950s. £100 £50 £20

**Space Gun (MTU Toys, Korea)**
Mainly blue with streamlined top back of weapon, screen
through which sparks fly, silver trigger and criss-cross
design in yellow, black, red, orange and white outline on
handle. The words 'Space Gun' in yellow on sides
alongside other neat designs and markings. Issued 1978. £50 £20 £5

**Space Laser Gun I (China)**
Battery operated silver gun with red bulb. Four sounds:
police, machine gun, ambulance and fire engine. In
colourful box. Price £3.95. £150 £50 £20

**Space Patrol Rocket Gun (Japan)**
Red with rockets. Very attractive box with the words
'Here is Your Official Space Patrol' in large black and red
lettering. Issued 1956. £100 £50 £20

**Space Patrol Rocket Fun (USA Plastic Co.)**
Red and white with yellow and red rockets. Space patrol
name on handle. Issued 1951. Limited edition. £50 £30 £10

**The Space Pistol (Palmer Toys)**
Almost all-black, bulb shaped water pistol with white
trigger. Stars on handle and trade mark. Issued 1967. £50 £20 £10

**Space Pistol Park Plastics**
Black, bulb shaped water pistol. Issued late 1950s or early
1960s. £50 £20 £10

**Space Pistol (Japan)**
One of the better water pistol guns in the form of an outer planet spaceman. All-copper or gold, bulb shaped body, with silver and red tinted trigger. Issued 1967/68.

£100 £50 £20

**The Space Pistol Potato Gun (Hong Kong)**
Black at rear handle part and red large trigger and barrel colour, although colours vary. Issued 1983/84.

£20 £10 £5

**Space Ray-Pop Gun (Wyandote All-Metal Products Co.)**
Red and brown. Issued 1935/38.

£200 £75 £30

**Sparker Gun (Hero Toy Co., Japan)**
Mainly yellow or gold with exotic dotted criss-cross design, red, blue and green high quality decor and silver trigger. Issued 1968.

£100 £40 £10

**Speedepth Space Rifle (Japan)**
Gold and black with picture design and markings on handle. Issued 1969. Limited edition. Rare.

£1000 £200 £50

**Star Trek Water Pistol (Aviva Enterprises, Taiwan)**
Silver with plastic cap and series of colour designs on top. Issued 1979.

£50 £20 £10

**Star Wars Space Attacker (Hong Kong)**
Large rifle-type weapon in red and gold, black and silver with pictures of men in spaceship along handle. Issued 1978. According to box, limited special.

£350 £100 £40

**Strato Weapon (Japan)**
High quality gun in all pure metallic silver with brown ring design in centre of weapon and on barrel tip and the word 'Strato' on handle. Issued 1965. Good investment.

£100 £50 £20

**Super Soaker Gun 20 (Larami Corp., Philadelphia, USA)**
Yellow with red trigger, spring end, top and lower barrel. Price £5.50. Issued 1991. Boxed.

£50 £25 £10

**Super Soaker Gun 30 (Larami Corp., Philadelphia, USA)**
Red with yellow, thick top barrel and white or silver lower barrel. Issued 1991. Price £6.95. Boxed.

£50 £25 £10

**Super Soaker Gun 50 (Larami Corp., Philadelphia, USA)**
Yellow and silver with green, thick top barrel. Issued 1991. Price £6.95. Boxed.

£50 £25 £10

| GUNS | MB | MU | GC |
|---|---|---|---|

**Super Soaker Gun 100 (Larami Corp., Philadelphia, USA)**
Blue with yellow top barrel and silver lower barrel. Issued 1991. Price £6.95. Boxed. — £75 £35 £15

| | £75 | £35 | £15 |

**Super Soaker Gun 200 (Larami Corp., Philadelphia, USA)**
Red and yellow with green top twin barrels and silver lower barrel. Issued 1991. Price £9.95. Superior model with special box and leaflet.

| | £100 | £50 | £20 |

**Super Soaker XTC (Larami Corp., Philadelphia, USA)**
Red and yellow with green twin barrels. Issued 1991. Boxed.

| | £50 | £20 | £10 |

**Supersonic Space Gun (Daiya Toys, Japan)**
Silver with red, brown and black design. Rather striking picture of rockets and spaceman on planet painted on sides of handle and body. Issued 1968.

| | £200 | £50 | £20 |

**The TG-105 Space Pistol (Hong Kong)**
Silver-grey. Black handle with the number '105' near trigger. Issued 1989.

| | £50 | £20 | £10 |

**The 3-Colour Space-Ray Gun (Ideal Toys)**
Red and blue. When trigger is activated three different colours flash out. Issued 1950s. Limited edition.

| | £200 | £50 | £20 |

**Tom Corbett Atomic Rifle (Marx Toys, New York, USA)**
Silver with the words 'Space Cadet' in red and a touch of lemon. Issued 1955.

| | £100 | £50 | £20 |

**The Tom Corbett Flashlight Pistol (Marx Toys, New York, USA)**
Red with blue trigger and white, clear plastic screw-type end. Colourful box. Issued 1956.

| | £50 | £30 | £10 |

**The Tom Corbett Gun (Marx Toys, New York, USA)**
Very thin weapon with blue barrel, with the words 'Space Cadet' along sides. Red and orange striped body with the words 'Space Gun' on black background and picture of spaceman on handle. Issue 1954.

| | £75 | £35 | £15 |

**The 238 Clicker Space Gun (Korea)**
Silver and black with yellow or red trigger, with spaceman picture on handle and the number '238' in bright yellow on white background. Other colours exist and value about the same. Issued 1989.

| | £40 | £20 | £10 |

**Viper Sting Gun (China or Japan)**
This is a rare model made in a limited edition which I
found at an exhibition at Earls Court, London in 1965.
Although I am fairly certain it was manufactured in
Japan, it could have been made in China. Silver and blue
with picture of Viper and his spaceship.

£500 £100 £50

**The White Ray Gun (Radio Shack, USA)**
All-white with red trigger and end bulb top section.
Well-designed model. Issued 1987/89.

£50 £20 £10

**The Zooka-Pop Pistol (Daisy, Plymouth, USA)**
Blue, red and orange with wording and decals, with
sound. Issued 1936.

£100 £50 £20

# OTHER SPACE
# SPECIALS

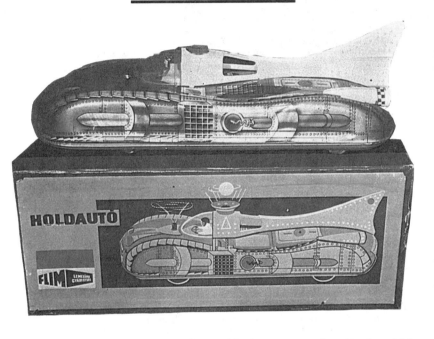

Listed below are space toys manufactured by important makers but for which there are not enough to justify arrangement under a separate company heading. Some of the manufacturers featured here are Action Man, Century 21, Majorette and Triang-Hornby. I have also included at the end some one-off models made by clever inventors.

| MODEL | MB | MU | GC |
|---|---|---|---|
| **Action Man No. 34701 Astronaut Set** | | | |
| Authentic astronaut's silver space suit with gloves, boots, space helmet, tether cord, oxygen chest pack, propellant gun and camera. Instruction leaflet. Price £1.25. Issued 1964. Deleted 1970. | | | |
| Without action man | £100 | £40 | £20 |
| With action man | £250 | £75 | £40 |

| | MB | MU | GC |
|---|---|---|---|
| **Action Man No. 34705 Space Capsule**<br>Floating space capsule, sliding canopy, control panel and retro-pack. Instruction leaflet. Made by Hassenfield Brothers, USA. Price £2. Issued 1964. Deleted 1970. | £200 | £75 | £30 |
| **Aladden Industries, Tom Corbett Thermos Flask**<br>Flask with the words 'Space Cadet' in large yellow or gold letters. Well designed with pictures of rocket and spaceman and woman in dark grey and blue space dress. Red cup. Nice box and packing. Issued 1950s. For several years this rare item was classed as unimportant as a collector's piece, and therefore many were thrown away. | £100 | — | — |
| **AMF-WEN-MAC, Craig Breedlove's Spirit of America**<br>This streamlined vehicle was for promotional business and available only through Shell and Goodyear. Authentic livery, clockwork motor and working brake parachute. The actual car itself broke the American speed record at 52628mph. Issued 1950. Deleted almost at once. Very rare. | £1000 | — | — |
| **Auburn Rubber Products, Alligator Rex**<br>This unusual model from a toy company in Indiana and Pennsylvania, USA would surely only be seen on some far away planet. In the shape of a beast in the alligator class with mouth wide open. Called a 'Green Unique', in yellow, red and black with silver trim. Issued 1936. Deleted almost at once. 190mm long. Very rare. | £750 | — | — |
| **Budgie No. 272 Supercar**<br>One of the first British diecast space toys. Red and silver with retractable wings and clear plastic canopy for driver/pilot. Price 5/6, although price varied. Issued shortly after the first few episodes of the TV series. | £1000 | £250 | £50 |
| **Century 21, Lady Penelope's Car**<br>Pink with silver wheels and white tyres and with silver radiator and bumpers. Price £5. Issued 1965. Deleted 1969. | £150 | £75 | £40 |
| **Century 21, Thunderbird 1**<br>Authentic livery and decals. Price £5. Issued 1965. Deleted 1969. | £250 | £100 | £50 |
| **Century 21, Thunderbird 2**<br>Authentic livery and decals. Price £5. Issued 1965. Deleted 1969. | £250 | £100 | £50 |
| **Century 21, Thunderbird 3**<br>Authentic livery and decals. Price £5. Issued 1965. Deleted 1969. | £250 | £100 | £50 |

**Century 21, Thunderbird 4**
Authentic livery and decals. Price £5. Issued 1965.
Deleted 1969.

|  | MB | MU | GC |
|---|---|---|---|
|  | £250 | £100 | £50 |

**Space Moon Buggy Playset No. 40604 by Schleich**
Far-away in a land where houses are shaped like
mushrooms live a community of 'blue people'. Many
think that smurfs dropped out of the mushrooms, while
others think they smurfed from nowhere. They were born
in 1957 from a cartoon strip by Professor Peyo, and made
as prototypes by a German company called Schleich.
Almost every kind of smurf exists, from the scholar, the
thinker, the doctor and even the cowboy, as well as a
community of space creatures named 'Astros'. The
creatures were sold by the National Benzole Petrol
Company in the 1960s and 1970s. They were also made
in Germany, Holland, Hong Kong and Portugal. The
Moon Buggy Playset was issued between 1983 and 1985,
and includes a Space Astro Smurf driving the moon
buggy pulling a trailer, a rocket ship flying the American
flag, and other accessories. Rare.

|  | MB | MU | GC |
|---|---|---|---|
| Playset | £1000 | £250 | £50 |
| Empty box | £400 | — | — |

**Coronet, Smurf Product No. 30  Astro Spaceman**
Also numbered 5341 and 20003. Clear plastic helmet.
Issued free with petrol 1973. Deleted 1980. Unboxed.

|  | MB | MU | GC |
|---|---|---|---|
| White suit | — | £10 | £2 |
| Cream suit and blue hands | — | £25 | £5 |
| Dark blue suit (very rare) | — | £500 | £50 |

**Ertle, Shuttle Model Special**
White and silver with removable space lab. Price £5,
although price varied. Issued 1968. Deleted 1976. 120mm
long. Much sought after high quality model.

|  | MB | MU | GC |
|---|---|---|---|
|  | £150 | £75 | £35 |

**Ertle, Shuttle with Diecast Booster Rockets**
White and silver with authentic space shuttle markings.
The rockets detach from the plastic fuel cell, which in turn
detaches from the diecast shuttle. The Taj Mahal effect is
noticeable. Price £9, although price varied. Issued 1968.
Deleted 1970. 76mm long.

|  | MB | MU | GC |
|---|---|---|---|
|  | £200 | £100 | £50 |

**Eurauto, Renault Radar Truck**
No. 3/98 with plastic back of No. 3/96 and tinplate radar
scanner. Olive matt. Price £2. Issued 1964. Deleted 1969.
121mm.

|  | MB | MU | GC |
|---|---|---|---|
|  | £100 | £40 | £20 |

| | MB | MU | GC |
|---|---|---|---|
| **Hannatoys, The Space Car Special**<br>Green and lemon two-tone effect with clear plastic cockpit cover. Much sought after streamlined model from the USA. Issued 1950s. Limited edition. | £100 | £50 | £10 |
| **Holdauto & Flim, Spacecraft**<br>Well-designed battery operated model from Hungary with curves and marks and with spaceman in plastic see-through cockpit. Rare box with the words 'Holdauto & Flim' on front. Issued 1960/65. Rare. | £300 | £150 | £75 |
| **Hong Kong Toys, Pencil Top Batman**<br>Green, blue, red or yellow, although colours vary. Not individually boxed. Price 5p. Issued 1966/77. | — | £10 | £5 |
| **Hong Kong Toys, Pencil Top Robin**<br>In an assortment of liveries to match the figure of Batman. Not individually boxed. Price 5p. Issued 1966/77. | — | £10 | £5 |
| **Hong Kong Toys, Pencil Top Space Creature**<br>These rubber models, made in Far Eastern countries, were sold without boxes, and were normally acquired by children who collected rubbers between 1966 and 1976, when the craze was at its height. They sold in their millions, and normally were 1 to 3in. in height like the standard sizes of rubbers. Prices ranged from only a few pence. The model here was in silver with a blue head and shoes and with antennae. | — | £20 | £10 |
| **JR Enterprises, Space Ranger Badge**<br>Gold with the words 'Space Ranger' in silver and the space ring outline attached to the wing designed badge. Issued 1954/58. | £50 | £20 | £10 |
| **Lintoy, Boeing Carrying Shuttle**<br>White and silver with exhausts cowled during test flights. Price £7, although price varied. Issued 1976. Deleted 1983. 90mm long. | £100 | £50 | £25 |
| **Lone Star, Batman Game**<br>Game of hoop-la in colourful design and box approx. 11in. square. Issued 1965/75. | £50 | £30 | £15 |

**Majorette, Extranimals Series 2100**
Majorette of France is one of the leading makers of super models, selling in over 70 countries. The beginning of the 23rd century saw the appearance of strange vehicles, half-vehicle, half-animal, called 'Extranimals'. No one knew where they came from, but everyone feared them – ferocious, powerful and fast over any obstacle. They kept their animal instinct and the power of the vehicles

in which they took form. Nothing can resist them on their planet of the future, as they have been made quite invincible by their incredible mutation. They fight for their survival on the place where they rule. Good investments for people who may have them and not realise their worth.

| | MB | MU | GC |
|---|---|---|---|
| **No. 2101 Mustang**<br>Red with silver tail and head. White plastic wheels and large black tyres with eight different wheel positions. Colourful box. Price £3.95, although price varied. Issued 1989. Deleted 1993. 260 x 160mm. | £50 | £20 | £10 |
| **No. 2102 Panther**<br>Red, grey, white and silver. White plastic wheels and thick black tyres with 8 different wheel positions. Price £3.95, although price varied. Issued 1989. Deleted 1993. 260 x 160mm. | £50 | £20 | £10 |
| **No. 2103 Taurus**<br>Blue, silver and white with yellow trim line. White plastic wheels and thick black tyres with eight different wheel positions. Price £3.95, although price varied. Issued 1989. Deleted 1993. 260 x 160mm. | £50 | £20 | £10 |
| **No. 2104 Rhino**<br>Gold and silver. White plastic wheels and large black tyres with eight different wheel positions. Price £3.95, although price varied. Issued 1989. Deleted 1993. 260 x 160mm. | £50 | £20 | £10 |
| **No. 2105 Scorpio**<br>Red or pink with silver tint. White plastic wheels and thick black tyres with eight different wheel positions. Price £3.85, although price varied. Issued 1989. Deleted 1993. 260 x 160mm. | £50 | £20 | £10 |
| **No. 2106 Eagle**<br>Fine unusual model. Red, silver, blue and black with silver prop. Price £3.95. Issued 1989. Deleted 1993. 260 x 160mm. | £60 | £30 | £15 |
| **No. 2130 Shark**<br>Orange, blue and silver. Six white plastic wheels and large thick black tyres. Price £4.75, although price varied. Issued 1989. Deleted 1993. 200 x 100 x 100mm. | £75 | £40 | £20 |

### No. 2131  Naja
Green, lemon and red. Six white plastic wheels and large thick black tyres. Front and back ramps. Price £4.75, although price varied. Issued 1989. Deleted 1993. 200 x 100 x 100mm.                                         £75    £40    £20

### No. 2150  Elephant
Dark blue or silver with red and silver. Large white plastic wheels and thick black tyres. Price £6.50, although price varied. Issued 1989. Deleted 1993. 270 x 130 x 160mm.                                                £100   £50    £25

### Majorette No. 610  Semi-Rocket Transporter
White and silver with red wheels, decals and markings. Rocket on rear. Price £8.95, although price varied. Issued 1989. Deleted 1993. 260 x 410 x 790mm.            £100   £50    £25

### Majorette, No. 610A  Rocket
Separate rocket in white and silver; a good standby in case something happened to the rocket on the transporter. Price £2. Issued 1989. Deleted 1993.     £50    £20    £10

### Majorette, No. 611  The Spacecraft Transporter
White with silver wheels, bumpers, grille, lights with red and yellow decals, and space wording and numbers. Rocket on back of transporter in white, red and yellow. Price £6.50. Issued 1989. Deleted 1993. 160mm long.   £100   £50    £25

### Marx Toys, Buck Rogers Rocket Police Patrol
Silver green with red nose with black and orange rings and with gold, orange and black eyes. Red rocket boosters underneath and the rib-cage design in red at rear. The figure of Buck Rogers in the driver's seat with weapon and silver collar opening of cockpit. The words 'Rocket Police Patrol' in yellow, red and black and 'Buck Rogers' in black with red and cream background. Issued 1935/40 in USA, although a chance of this model turning up after the war was a possibility. Very rare, probably impossible to find in mint boxed or even mint unboxed condition.     £500   £200   £100

### Marx Toys, Buck Rogers Cast Iron Rocket
Dark green with red and white markings on solid iron base. Issued 1938/39. No known box for model apart from a plain type of carton made from gunpaper card with a number attached. A prewar treasure which made history and provided the basis for many more space models to come.                                           £500   £100   £50

**Marx Toys, The Dalek**
Authentic livery and logos. Colourful box. Issued
1960/70. Good investment.

| | MB | MU | GC |
|---|---|---|---|
| Marx Toys, The Dalek | £200 | £100 | £50 |

**Milton Bradley, Captain Video Space Game**
This well-designed space game came in a colourful
presentation box in blue, gold and black with red border.
The words 'Captain Video Space Game' in light blue and
'Captain Video' in white letters. Issued 1950s. £50 £20 £10

**Politoy No. 129  Spyder Car**
Politoy, or Polisti of Italy, made this rare model with the
serial number '89' stamped into the chassis. Lavender
with metallic blue interior, with a very attractive
presentation box showing the car and a pretty girl on the
front. Replica of a car driven by the Italian entertainer and
space-crazy personality, Rita Pavone. A red-haired
charmer, she appears in the picture on the box wearing a
blue dress and straw hat. On the back of the picture was
a list of her records by RCA plus a handwritten message
in Italian. Price £5. Issued 1970. Deleted 1976.

| | MB | MU | GC |
|---|---|---|---|
| Unsigned | £100 | — | — |
| Signed | £1000 | — | — |
| Unsigned empty box | £250 | — | — |
| Signed empty box | £1000 | — | — |

**Saviem No. 3/97  Veronique Rocket Launcher**
This very unlikely vehicle is painted in blue with a black
chassis. The tinplate trailer is made on the front unit of
No. 3/75 and has a tinplate swivelling ramp with a crude
spring operated firing mechanism. A rocket aimer sits on
the operator's seat and wears oilskins or a spacesuit in
bright yellow. Hollow red plastic rocket with two bands
of chrome tape. Price 10/6, although price varied. Issued
1964. Deleted 1970. Rocket 160mm long. Overall model
225mm long. £250 £75 £40

**Solido No. 2201  Technical Missile Launcher**
Green with black tyres and red rocket missile. Price £9,
although price varied. Issued 1980/81. Deleted 1986. £250 £75 £35

**Solido No. 3810  Europe Assistance Helicopter**
White with black prop, tail and skis. Tinted glass to see
into cockpit with black markings and orange
background. Price £7, although price varied. Issued 1981.
Deleted 1986. £250 £100 £50

**Triang-Hornby No. RS16  Strike Force 10 Set**
Mobile air and ground combat train. Contains No. R562
Catapult Plane Launch Car and No. R568 Assault Tank
Transporter hauled by a synchrosmoke locomotive. Dark
green with orange tint lines and orange and white decals.
Close-support plane is launched when railcar passes over
trackside trigger. Flies over 20ft. Tank has twin red-eye
rocket launchers. Turret has 360° traverse and variable
elevation. Items could be bought separately except
special locomotive. Price £12, although price varied.
Issued 1967. Deleted 1969.

£750 £250 £100

**Triang-Hornby No. R128K  Battle Rescue Helicopter Car**
Yellow and white with grey skis. As the railcar passes the
trackside trigger, the pre-wound launch mechanism sets
off the helicopter. Price £5. Issued 1967. Deleted almost at
once.

£200 £50 £30

**Triang-Hornby No. R239  Red Arrow Bomb Transporter**
Massive blockbuster bomb in red and silver nose. Cap
loaded warhead which detonates on hitting hard surface.
Price £4, although price varied. Issued 1967. Deleted 1970.

£200 £100 £40

**Triang-Hornby No. R249K  Exploding Car**
Dark green with orange dots and decor. Explodes in
realistic fashion when hit by a battle space rocket missile,
but can be easily reassembled. Price £5. Issued 1967.
Deleted 1973.

£350 £100 £50

**Triang-Hornby No. R343K  Multiple Missile Launcher**
Green, blue and dark green with orange dot and line
decor. Rotating armoured turret packs four spring loaded
rocket launchers. Hand triggers fire red-eye white and
black nose-tip rockets singly or in savage salvo. Price £7,
although price varied. Issued 1967. Deleted 1973.

£250 £50 £25

**Triang-Hornby No. R562  Catapult Launch Car**
Authentic livery and decals. Price £3. Issued 1966.
Deleted 1969.

£100 £50 £10

**Triang-Hornby No. R568  Assault Transporter**
Authentic livery and decals. Price £3. Issued 1966.
Deleted 1969.

£100 £50 £10

**Triang-Hornby No. R571  G-10 Q Car**
Dark green with 'G-10' in yellow. Freight wagon houses
two red-eye rockets on launchers. As the wagon passes
the trackside trigger, the roof and sides fall away to allow

| | MB | MU | GC |
|---|---|---|---|
| rocket launchers to swing round through 90° and elevate into firing position. Price £5, although price varied. Issued 1967. Deleted 1969. | £200 | £100 | £50 |

**Triang-Hornby No. R630  POW Car**
Transports prisoners of war away from combat zone and also carries freight and ammunition. Price £4. Issued 1967. Deleted 1970.

| | £50 | £25 | £10 |
|---|---|---|---|

**Triang-Hornby No. R631  Recovery Wagon**
Rail mounted heavy-duty crane for vehicle and goods recovery. Dark green with yellow decals and lettering. Price £5. Issued 1967. Deleted 1970.

| | £60 | £30 | £15 |
|---|---|---|---|

**Triang-Hornby No. R670  Twin Ground to Air Missile**
Green and dark green with white and black nose-tip rockets. Camouflaged fortified position. Turret rotates and launchers elevate for firing. Price £3.50, although price varied. Issued 1967. Deleted 1970.

| | £75 | £40 | £20 |
|---|---|---|---|

**Triang-Hornby No. R671 Multiple Ground to Air Missile Site**
Green and dark green with white and black nose-tip rockets. Radar controlled, red-eye rocket unit in rotating armoured turret and mounted in fortified position. Spring loaded launchers and fingertip firing. Price £5. Issued 1967. Deleted 1970.

| | £150 | £75 | £40 |
|---|---|---|---|

**Triang-Hornby No. R672  Honest John Pad**
Hand based missile site for defence with rubber warhead. Grey with black head and with orange and black square markings. Price £5, although price varied. Issued 1967. Deleted 1970.

| | £300 | £100 | £50 |
|---|---|---|---|

**Triang-Hornby No. R725  Command Car**
This battle group headquarters for staff officers picks up and delivers dispatches, maps, rations, ammunition, rockets and other material for any space war. Price £4, although price varied. Issued 1967. Deleted 1970.

| | £150 | £75 | £35 |
|---|---|---|---|

**Triang-Hornby No. R752  Battle Space Turbo Car**
Unusual model included in railways with its variable speed propeller-driven device. Red with silver trim and decals, slitlike windows at the nose and large white propeller at the rear. Ramming spike and four grey or black wheels. Price £3, although price varied. Issued 1967 in limited edition. Deleted almost at once.

| | £500 | £100 | £50 |
|---|---|---|---|

| | MB | MU | GC |
|---|---|---|---|
| **Triang-Hornby No. 0750  Medical Aid Car**<br>White and green with Red Cross decals on white background. Serves as mobile field hospital. Price £5. Issued 1967. Deleted 1970. | £100 | £50 | £20 |
| **Triang-Hornby No. 0751  Satellite Set**<br>Complete train with No. R566 Spy Satellite Launcher and No. R567 Radar Tracking Command Car hauled by locomotive. Pre-wound launch mechanism boosts spinning satellite into air on mission as railcar passes over trackside trigger. Revolving radar scanner on command car tracks satellite or enemy crafts or weapons. Flashing dome shows all-systems go. Price £10, although price varied. Issued 1967. Deleted 1970. Rare and much sought after investment. | £550 | £200 | £100 |
| **Triang-Hornby No. 0756  Tactical Rocket Launcher**<br>Green with powerful red and black rocket. Spring loaded system with the long-range missile, known as 'Big-X'. Price £5, although price varied. Issued 1967. Deleted almost at once. | £250 | £50 | £25 |
| **Vostok Toys, Satellite**<br>Fine model from Russia which put the first man in space and has always been a leading force in the outer planets. White. Price £12.50, although price varied. Issued 1982. Deleted 1989. | £30 | £15 | £10 |
| **Western Toys No. WMS30  Thunderbolt**<br>Silver. Streamlined as a futuristic model. Price £1.50. Issued 1976. Deleted 1983. Rare to find in either solid or kit form. | £250 | £75 | £45 |
| **The Reg Pollard NASA Space Shuttle**<br>Certainly one for the record books, a model built by matchstick fanatic Reg Pollard of Manchester, who lived and worked in a house on the outskirts of the city. In 1982 he completed the model, true to scale, from a total of 77735 matchsticks, taking 1307 hours. 54in. long. Definitely difficult to value. | £750 | — | — |

**Taylor Aerocar Flyer**
This model was made as a promotional gimmick in 1948 by Gladden Plastics for Aerocar Corporation, Longview, Washington, USA from a vehicle designed and built by a former naval commander, Moulton B. Taylor. The vehicle received its certificate of air worthiness in 1956. However, production of the model ceased in 1960 after only a total of 7 or 8 units had been produced over the whole 12 years. People were not quite ready for space in a flying car and, furthermore, the helicopter was appearing successfully.

The model was also made in plastic kit form. Original price not certain as models in both forms were given away for promotional purposes, although some were sold privately as purely a collector's item and investment. 1/25th scale.

| | MB | MU | GC |
|---|---|---|---|
| Model | £2000 | — | — |
| Kit form | £4000 | — | — |

**The Taylor Made Aerocar**
This was a design exercise model of a rear-engined three-wheeler and tail-fin 39 Lincoln sedan wind tunnel vehicle. Authentic Taylor decals and livery. Another promotional model which would have been an exclusive for only a few selected people. Model built possibly 50 years ahead of its time. This type of car could only really be at home on a planet in outer space. Issued and deleted 1945.

| | MB | MU | GC |
|---|---|---|---|
| | £750 | — | — |

# DOCTOR WHO

Who would have thought that this series could have gained so much popularity in every corner of the world? Now it will be possible to enjoy the Daleks and the Cybermen for ever since the episodes are available on audiotapes and videos!

Doctor Who has been played on television by a number of actors since the series started in 1963: to date, they have been William Hartnell, Patrick Troughton, Jon Pertwee, Tom Baker, Peter Davison, Colin Baker and Sylvester McCoy. (In addition, Peter Cushing played the part in two films.) Any of the following audiotapes or videos with the dated signature of the actor playing the part of Doctor Who is worth anything between £1,000 and £10,000, and it is advisable to consult an expert or take legal advice in such instances. While unsigned audiotapes and videos are reasonably priced at the present time, they will rise in value and become very good investments.

|  | Mint | Used |
|---|---|---|
| **AUDIOTAPES** | | |
| **Ark in Space, No. DTO 10519**. Price £1.50. Taken from video. Rare. | £50 | £10 |
| **Evil of the Daleks, No. 32996**. Price £7.99. | £10 | £2 |
| **Genesis of the Daleks, No. 62948**. Price £7.99. | £10 | £2 |
| **The Macra Terror, No. 32997**. Price £7.99. | £10 | £2 |
| **Power of the Daleks, No. 52629**. Price £7.99. | £10 | £2 |
| **State of Decay, No. DTO 10517**. Special release of a two cassette pack by Pickwick International. Price 99p. Rare. Signed by Tom Baker | £25 £2000 | £5 — |
| **Tomb of the Cybermen, No. 52751**. Price £7.99. | £10 | £2 |
| **VIDEOS** | | |
| **Ark in Space, No. 60615**. Price £10.99. | £15 | £5 |
| **Aztecs, No. 36582**. Price £10.99. | £15 | £5 |
| **Brain of Morbius, No. 22873**. Price £10.99. | £15 | £5 |
| **Castrovalva, No. 26440**. Price £10.99. | £15 | £5 |

| DOCTOR WHO | Mint | Used |
|---|---|---|
| **Caves of Androzani, No. 25095**. Price £10.99. | £15 | £5 |
| **City of Death, No. 91405**. Price £10.99. | £15 | £5 |
| **Claws of Axos, No. 30448**. Price £10.99. | £15 | £5 |
| **Curse of Fenric, No. 77337**. Price £10.99. | £15 | £5 |
| **Cybermen, the Early Years, No. 32660**. Price £12.99. | £20 | £8 |
| **The Daemons, No. 39946**. Price £10.99. | £15 | £5 |
| **Dalek Invasion Parts 1 and 2, No. 68678**. Price £19.99. | £30 | £10 |
| **Daleks Dead Planet/Expedition, No. 60649**. Price £19.99. Black and white. | £50 | £10 |
| **Daleks, the Early Years, No. 32601**. Price £12.99. | £20 | £8 |
| **Day of the Daleks, No. 22864**. Price £10.99. | £15 | £5 |
| **Deadly Assassin, No. 97129**. Price £10.99. | £15 | £5 |
| **Death to the Daleks, No. 73526**. Price £10.99. | £15 | £5 |
| **Doctor Who 30th Anniversary 1963/93, Nos. 52688, 54429, 46726 and 46681**. With John Pertwee discovering Silurians in Derbyshire and unravelling the feared curse of Peladon. With Patrick Troughton in 'The Invasion', investigating the sinister firm International Electromatics. With Tom Baker, struggling to save the sacred 'Keeper of Traken'. Price of set £56. | £75 | £50 |
| **Dominators, No. 74595**. Price £10.99. | £15 | £5 |
| **Earthshock, No. 34496**. Price £10.99. | £15 | £5 |
| **Enlightenment, No. 39259**. Price £10.99. | £15 | £5 |
| **Five Doctors, No. 71381**. Price £10.99. | £15 | £5 |
| **Hartnell Years, No. 92192**. Black and white. Price £10.99. | £25 | £10 |
| **Image of the Fendahl, No. 39945**. Price £10.99. | £15 | £5 |
| **The Invasion, No. 46726**. Price £10.99. | £15 | £5 |
| **Keeper of Traken, No. 46681**. Price £10.99. | £15 | £5 |
| **Krotons, No. 22876**. Price £10.99. | £15 | £5 |
| **Logopolis, No. 26441**. Price £10.99. | £15 | £5 |

| | Mint | Used |
|---|---|---|
| **Masque of Mandragora, No. 36111**. Price £10.99. | £15 | £5 |
| **Mawdryn Undead, No. 36314**. Price £10.99. | £15 | £5 |
| **Mind Robber, No. 22872**. Price £10.99. | £15 | £5 |
| **Pertwee Years, No. 26520**. Price £10.99. | £20 | £8 |
| **Planet of Spiders, No. 91404**. Price £19.99. | £25 | £10 |
| **Power of the Daleks (Special Doctor Who), No. 52629.** With Patrick Troughton. Price £7.99. | £10 | £3 |
| **Pyramid of Mars, No. 22859**. Price £10.99. | £15 | £5 |
| **Revenge of Cybermen, No. 22852**. Price £10.99. | £15 | £5 |
| **Robot, No. 25094**. Price £10.99. | £20 | £8 |
| **Robots of Death, No. 51065**. Price £10.99. | £20 | £8 |
| **Seeds of Death, No. 22860**. Price £10.99. | £15 | £5 |
| **Shada, No. 32602**. Price £10.99. | £15 | £5 |
| **Silver Nemesis, No. 41250**. Price £10.99. | £15 | £5 |
| **Sontaran Experiment/Genesis of the Daleks, No. 97142.** Price £19.99. | £25 | £10 |
| **Spearhead from Space, No. 51066**. Price £19.99. | £15 | £5 |
| **Talons of Eng-Chiang, No. 59641**. Price £19.99. | £15 | £5 |
| **Terminus, No. 38929**. Price £19.99. | £15 | £5 |
| **Terror of the Autons, No. 41252**. Price £19.99. | £15 | £5 |
| **Terror of Zygons, No. 59642**. Price £19.99. | £15 | £5 |
| **Three Doctors, No. 93884**. Price £10.99. | £15 | £5 |
| **Time Warrior, No. 60616**. Price £10.99. | £15 | £5 |
| **Tom Baker Years, No. 34487**. Price £10.99. | £15 | £5 |
| **Tomb of the Cybermen, No. 54195**. Price £12.99. | £20 | £8 |
| **Troughton Years, No. 92193.** Colour/black and white. Price £10.99. | £25 | £10 |
| **Vengeance on Varos, No. 42838**. Price £10.99. | £15 | £5 |

**War Games Parts 1 and 2, No. 22871**. Black and white. Price £19.99. — £30 — £10

**Web Planet Parts 1 and 2, No. 22875**. Price £19.99. — £30 — £10

## COMPACT DISC

**30 Years at the Radiophonic Workshop, No. 54284**. Over 80 extraterrestrial sound effects and signature tunes that helped to make Doctor Who compulsory viewing for so many. Price £10.99. Rare one-off issue. — £25 — £10

# STAR TREK

I have been interested in space toys and models ever since I saw the films featuring Flash Gordon, played by Larry Buster Crabbe.

Many familiar actors names come to mind in the world beyond the stars: Raymond Massey, Lorne Greene, William Shatner, DeForest Kelley, Michael Ansara, Leonard Nimoy and James (Scottie) Doohan – to name but a few.

Then there are the writers. H.G. Wells gave us a future world of space in *The Shape of Things to Come*. Gene Roddenberry is one of the greatest names in the history of spaces. Trained in aeronautical engineering and a former airline pilot, he combined these with his skills as a writer, which provided the perfect balance for the creation of that unique science fiction institution in the world of television, Star Trek.

Star Trek set the world crazy on collecting robots, airships and space modules as well as badges, buttons and posters. It came out of two things: the reading of science fiction and a censored television, when no one was allowed to talk about sex, politics, religion or war of any kind!

Roddenberry decided to steal a page or two from Jonathan Swift's *Gulliver's Travels*, and introduce some strange 'Polka-dotted People' and some things that the censors would overlook. He was right. He was helped by Desilu Productions and its executive, the actress Lucille Ball, who gave Star Trek the start it needed. By the time the series was sold to Paramount at the time of its fiftieth episode, the world had gone mad on Star Trek.

It hardly seems more than a week ago when I saw the first episode. From the start, writers like George Clayton Johnson, Dorothy Fontana, Samuel Peeples, John S.D. Black and Richard Matheson were hired, making certain that this series was going to be the best there had ever been of its kind. This was in the late 1960s, and the series is still running today. In the first year, 78 programmes lasting 53 minutes each were shown on the TV screens of America. When the series was shown 5 times a week on 142 US TV stations and, in addition, 22 animated episodes began a 2-season run in 1972, the biggest boost of all times was given to the collecting of space toys.

When it comes to collecting and investment, it is always useful to know what to look out for, especially when it comes to items connected with Star Trek, Star Wars and the like. If there is a well-known personality connected with the product, try and obtain an autograph. And remember: get the star to date his or her signature since it is important to do so.

## EPISODES 1–78

If any of the boxes for the following videos is signed by a star of the episode then it is worth £1000.

### Pilot  The Cage, by Gene Roddenberry
The original pilot episode for the series was turned down by NBC for not enough action, and for a long time it was believed that the colour footage had been lost or destroyed. Then, in 1988, Paramount Pictures was able to recreate the episode in its original form. This episode was never shown in its entirety on TV, although parts were incorporated in the two-part episode, 'The Menagerie' (see No. 11).

No doubt millions of people still believe that Captain Kirk (played by William Shatner) was the original captain of the Starship Enterprise, but Kirk was immediately preceded by Captain Christopher Pike (played by Jeffrey Hunter). On the first voyage of the Enterprise, Pike tries to rescue an Earth crew that had disappeared 18 years earlier. However, this leads him into a trap and he is imprisoned in a cage where he is studied by a Higher Life Form. In addition to Jeffrey Hunter, the stars of this 64-minute, colour episode are Susan Oliver, Leonard Nimoy, Majel Barrett, John Hoyt, Peter Duryea and Laurel Goodwin. Price £10.99. Issued in a limited edition by BBC Direct Sales as part of a set of three videos along with Star Trek 1–5 and Star Trek 2 and 3 (also priced £10.99 each;  the set could be bought at a special price of £29.99).

| | Mint | Used |
|---|---|---|
| No. VHR 2373  The Cage | £15 | — |
| Signed by Jeffrey Hunter or Gene Roddenberry | £50000 | — |
| Signed by another member of the cast | £10000 | — |
| No. D7397  Star Trek 1–5 | £15 | — |
| No. D7398  Star Trek 2 and 3 | £15 | — |
| Either of the above signed by William Shatner or Gene Roddenberry | £10000 | — |
| Either of the above signed by another member of the cast | £2000 | — |

### No. 1  The Man Trap, by George Clayton Johnson
There were as many candidates for the lead episode of Star Trek as there were fans, or at least stories. This episode featured a fat part for DeForest Kelley as McCoy. Jeanny Bal plays an old girlfriend of McCoy's, Francine Pyne the monster.

| | Mint | Used |
|---|---|---|
| | £50 | £20 |

**No. 2  Charlie X, by Dorothy Fontana**
Robert Walker Jr was the guest star in the title role, that of an adolescent raised in a world inhabited by Noncorporeal Beings. This episode deals with his infatuation with his life among humans and with his crush on Yeomand Rand. He was a dangerous character and this episode has a very poignant ending.    £50    £20

**No. 3  Where No Man has Gone Before, by Samuel Peeples**
This episode was originally produced as the second pilot for the series. Garry Lockwood and Sally Kellerman starred, the latter little realised that she was to set the world on fire as 'Hotlips' in Robert Altman's M\*A\*S\*H.    £50    £20

**No. 4  The Naked Time, by John S.D. Black**
Part of the interest and fun of any continuing series is watching characters develop week by week. Of course Star Trek has its competitors. However, not only are its chief characters intrinsically more interesting than most, but also its futuristic setting allows for some radical changes as well as gradual development. In this episode the Enterprise is crashing into a disintegrating planet, and McCoy has to find an antidote ... and fast.    £50    £20

**No. 5  The Enemy Within, by Richard Matheson**
One of the more astute creations made during the construction stage of the series was surely the Transporter Device. In story-telling terms alone (eliminating excess and then-he-went-to-time) it is invaluable. Besides, the transporter is a universal dream. Here Kirk is beamed aboard as two different people, the good Kirk and the bad Kirk, both in danger of dying from separation.    £50    £20

**No. 6  Mudd's Women, by Stephen Kandel**
Three miners are holding out on a much needed Dilithium Crystal discovery, and the trade they have in mind is more than Kirk's conscience can take. Harcourt Fenton Mudd is played by Roger C. Carmel, who became a continuing character.    £50    £20

**No. 7  What are Little Girls Made of?, by Robert Block**
An android is generally defined (though you would have to get a fairly recent Webster's Dictionary to find it listed at all) as an 'automaton' or 'robot' which resembles a human. In this story of the Mad Dr Corby, played by Michael Strong, his androids are set to take over the Enterprise.    £50    £20

### No. 8  Miri, by Adrian Spies

Several times in the series, the producers seem to have looked rather askance at youth: this first-season offering makes the children an anarchic group of vicious beasts. It seems that a fatal disease (the first symptoms are like leprosy) has killed all the adults and also kills the children as they reach puberty, an event occurring after a life span of hundreds of years. When the Enterprise crew beam down, they too begin the disease process, but they find a cure for themselves and the children. Kim Darby of the film True Grit fame starred in this episode.

£50  £20

### No. 9  Dagger of the Mind, by Simon Wincelberg

People seeing this episode only in syndication may forget that its initial airing and story must have had an impact on the then raging debate about psychosurgery, the radical techniques available for the control of mental patients. Dr Adams, the director of the penal colony on Tantalus, has devised a Neural Neutraliser in order to keep his charges in line. James Gregory has a high time as Adams, a role somewhat similar to that played by Charles Laughton in Island of Lost Souls, a film version of H.G. Wells's classic story, *The Island of Doctor Moreau*.

£50  £20

### No. 10  The Corbomite Maneuver, by Jerry Sohl

A rotating cube, turning on a corner like a giant multicoloured die, shows up in front of the Enterprise. Eventually, Kirk orders it to be destroyed, only to be faced with a larger problem (in more ways than one). Clint Howard is the guest star in this episode.

£50  £20

### No. 11  The Menagerie, by Gene Roddenberry

Here, in a two-part episode, footage from the original pilot episode for the series is used as a flashback within a story, which has Spock quite off the wall. Guest stars are Susan Oliver, Julie Parrish and Jeffrey Hunter.

£50  £20

### No. 12  The Conscience of a King, by Barry Trivers

This episode owes more than its title to Shakespeare's *Hamlet*. There is a mass murderer known as Kudos the Executioner.

£50  £20

### No. 13  Balance of Terror, by Paul Schneider

We have seen in our own time the speed with which each new weapon is made obsolete by a successive technological breakthrough. In this story, the Romulans have developed new weaponry (and tested it on Federation Outposts), as well as a screening device that makes their ships invisible to the outposts. Mark Leonard stars as the Romulan commander.

£50  £20

**No. 14 Shore Leave, by Theodore Sturgeon**
Shades or premonitions of Westworld; only the crew of
the Enterprise fail to be entertained by this amusement
park where everything you could think of comes true.
Locations filmed in Africa, near Los Angeles.     £50    £20

**No. 15 The Galileo Seven, by Oliver Crawford and S.
Bar-David**
Spock and other crew members set out in the shuttlecraft
Galileo to investigate the Quaser, Murasaki 312. They
must set down for repairs on a nearby planet, which is
occupied by a race of Goliath-like monsters, and no help
is possible. This is Spock's first command situation, faced
with mutiny. Guest stars are Don Marshal, Meter Marko,
Grant Woods and Rees Vaughn.     £50    £20

**No. 16 The Squire of Gothos, by Paul Schneider**
A sort of Liberace of space, Trelane is quite the snappy
dresser – velvet waistcoat, lace ruffles at the neck, the
works. William Campbell plays Trelane.     £50    £20

**No. 17 Arena, by Gene Coon and Fredric Brown**
By and large, Star Trek managed to avoid the
monster-of-the-week syndrome with which TV tends to
infect its sci-fi offerings. But the crew of the Enterprise
did have to deal with monsters from time to time, and
this episode had one of the best. Reminiscent of the
Creature of the Black Lagoon, the Gorn, captain of an
enemy ship, is matched with Kirk for a battle unto death.     £50    £20

**No. 18 Tomorrow is Yesterday, by Dorothy Fontana**
For those who do (or want to) believe in UFOs, this
episode offers the answer. The Enterprise itself, through
a time warp, finds itself tracked in the skies above
twentieth-century Earth. A pursuing jet is about to be
destroyed when Kirk orders the pilot beamed aboard in
order to save his life. Roger Perry is the guest pilot.     £50    £20

**No. 19 Court Martial, by Don Manliewicz and Stephen
Carabatos**
Elisha Cook Jr, perhaps best known as Sydney
Greenstreet's gunsel in John Huston's The Maltese
Falcon, turns in a typically perfect performance as a
lawyer with a reputation for pulling off hopeless cases. A
great classic episode.     £50    £20

**No. 20 The Return of the Archons, by Boris Sobelman**
This involved story takes on everything from
Communism to organised religion (and points out the
similarity between the two) in the context of an exciting
plot.     £50    £20

**No. 21  Space Seed, by Gene Coon and Carey Wilbur**
Our own immediate future appears during the series
only in bits and pieces. Here we learn something of the
Eugenics Wars that occupied (or will occupy) the last
twentieth decade of the century. Madlyn Rhue plays the
part of Khan Singh, the friend of a highly developed
creature played by Ricardo Montalban, of Planet of the
Apes fame.                                              £50      £20

**No. 22  A Taste of Armageddon, by Robert Hammer
and Gene Coon**
What at first appears to be a bloodless computer between
the planets Eminar II and Vendiker is found to be quite
deadly when the Enterprise is counted as a casualty. The
rules of the game insist that the crew submit themselves
to voluntary extermination, but what seems logical to the
planets in the situation is more than Kirk can take. There
are many stunning special effects in this episode.     £50      £20

**No. 23  This Side of Paradise, by Dorothy Fontana** (from
a story by Dorothy Fonatana and Nathan Butler)
As romantic as the F. Scott Fitzgerald novel from which
it takes its title, this story has the Enterprise crew
members falling over themselves and Spock falling in
love with the very beautiful Leila. She is played by Jill
Ireland (also known as Mrs Charles Bronson).           £50      £20

**No. 24  Devil in the Dark, by Gene Coon**
One of the most serious and mystifying questions of
space travel is our relationship with life forms different
from Earth's. This is a Beauty and the Beast story, and
guest stars include Janos Prohaska, Ken Lynch, Barry
Russo and Brad Weston.                                 £50      £20

**No. 25  Errand of Mercy, by Gene Coon**
Placed as they are between the Klingon Empire and the
Federation, the Organians are not really in a position to
be pacifists. Yet that is precisely the attitude they take in
this episode. The guest cast includes John Abbott, John
Colicos, Victo Lundin and Peter Brocco. These Organians
do not like violence at all.                           £50      £20

**No. 26  The Alternative Factor, by Don Ingalls**
An extension of the idea of 'The Enemy Within' episode
(and one that would continue in a second-season story),
here it is not Kirk who is divided in half but a space
traveller named Lazarus, played by Robert Brown.       £50      £20

**No. 27  City on the Edge of Forever, by Harlan Ellison**
Is history just a house of cards? Move one card, alter a
single moment of the past, and the entire structure
collapses. The question is an obsession with almost every
Star Trek writer, but has rarely been so interestingly or
tragically dealt with as in this story, which was honoured
by both the Hugo Awards and the Screen Writers Guild.
The plot involves a doctor (overdosed on cordrazine).
Joan Collins is a special guest, who holds the key to future
time.                                          £50     £20

**No. 28  Operation Annihilate, by Stephen W. Carabatos**
A glimpse, although under painful circumstances, of
Kirk's own family. His brother, together with his family,
live on Deneva, where they die from a parasite that
produces painful madness, then death. This final episode
of the first season features Craig Hundley (as Peter), Dave
Armstrong, Joan Swift and Maurishka Taliferro.       £50     £20

**No. 29  Amok Time, by Theodore Sturgeon**
This was the triumphant première of the second season,
when Spock lets his ears down in a most interesting (and
informative) episode. Discovered to be losing both his
skill and his cool, Spock reveals the secret passion of the
Vulcans. Guest stars include Arlene Martel (T'Pring) and
Celia Lovsky (T'Pau). The audiences came on tenfold
after this change of character (Ears Spock).          £50     £20

**No. 30  Who Mourns for Adonis?, by Gilbert Raiston**
One of the more unexpected episodes, where a giant
hand appears in the middle of space to stop the Enterprise
in its tracks (or treks). The hand belongs to Apollo of
Greek mythology fame.                         £50     £20

**No. 31  The Changeling, by John Meredith Lucas**
One of the most thoughtful and provocative episodes of
the series. The theme is the aftermath of man's creation.
Buy this and be amazed in a very special way.       £50     £20

**No. 32  Mirror Mirror, by Jerome Bixby**
With Spock at the controls, the transporter has a
near-fatal malfunction, beaming aboard four crewmen
(Kirk, McCoy, Scott and Uhura) who are not quite as they
appear. Instead, they are the alternate-universe
counterparts to the four. Barbara Luna guests as Marlene.    £50     £20

**No. 33  The Apple, by Max Ehrlich and Gene Coon**
This is from a great story. New sets on a weekly basis are
a tough order, although the Star Trek people always did
their best. However, the stone head of Vaal in this episode

bears an unfortunate resemblance to papier mâché. The story survives the problem. See this episode, buy it, enjoy the Apple of Eden.

£50     £20

**No. 34  The Doomsday Machine, by Norman Spinrad**
Within a decade after this episode was first aired, Space 1999 was to use a similar device for one of its stories: a Ship Graveyard, a sort of Bermuda Triangle on wheels, sucking in wayward craft. In this story feel the secret of the Doomsday Machine. Is it the real H-bomb of space? William Windom is the star who plays Commodore Decker.

£50     £20

**No. 35  Catspaw, by Robert Bloch and D.C. Fontana**
What seems to be voodoo first heats up the Enterprise to an extremely high temperature and then encases it like a fly in amber, and a giant black cat. One of the most supernatural of all the episodes. See this, then visit the space museum in Washington, DC if you can.

£50     £20

**No. 36  Mudd, by Stephen Kandel**
Can the irrepressible Harry Mudd really have met his end when sent off to prison? (See No. 6.) Did Batman ever completely rid himself of the Joker, or Superman of Luthor? No, Evil (like Hope) springs eternal; as do its more personable practitioners. Mudd is again played by Roger C. Carmel. Clever robots, great screening and a classic episode.

£50     £20

**No. 37  Metamorphosis, by Gene L. Coon**
A very beautiful and touching story – of love (however abnormal) and loneliness. Kirk, Spock, McCoy and a young woman diplomat for whom they are trying to get medical help are forced to land on a supposedly empty planet. There they discover a stranded space pioneer who has been on the planet for 200 years. Guest stars are Glenn Corbett and Elinor Donahue.

£50     £20

**No. 38  Journey to Babel, by Dorothy Fontana**
A conference is called to decide the thorny question of Coridan's admission into the Federation, and the Enterprise is given the task of transporting the bitterly divided delegates. Jane Wyatt, who appears as Spock's mother, Amanda, played Margaret in the previous series. Mark Leonard plays Spock's father, but was the Romulan commander in No. 13.

£50     £20

**No. 39  Friday's Child, by Dorothy Fontana**
On Capella IV, the Klingons stir up trouble by
encouraging a sympathiser to challenge the Teer (chief)
for leadership. Julie Newmar, Tige Andrews, Michael
Dante and Cal Bolder are the very special guests. £50 £20

**No. 40  The Deadly Years, by David Harmon**
Kirk loses command of the Enterprise, and much of his
sense. Commander Stocker, played by Charles Drake,
assumes command. The Romulans are involved in shots
from No. 13. £50 £20

**No. 41  Obsession, by Art Wallace**
A sentient cloud begins to pick off crew members in a
fashion disturbingly familiar to Kirk. Enveloping a
human in a few seconds, the cloud leeches all the red
blood cells from its victims' bodies. Stephen Brooks plays
the son of a dead captain, Garrovick. £50 £20

**No. 42  Wolf in the Fold, by Robert Bloch**
N R and R break for the Enterprise crew on a planet where
violence has been abolished almost turns into a violent
end for Mr Scott. First one and then another brutal
murder occurs, with Scott the prime suspect. The trail of
death continues, this time with the planet's High
Priestess being stabbed in Scott's arms. This has a Jack the
Ripper attached to it. The Exorcist also exists. £50 £20

**No. 43  The Trouble with Tribbles, by David Gerrold**
Well, it just seems like this show plays on TV weekly. For
a variety of reasons, this simple story has become one of
the most familiar of all the episodes, and the creature
called a tribble is recognised worldwide. The plot, which
involves a new grain hybrid and an uppity Federation
official, is quite secondary to the little fur creatures. £50 £20

**No. 44  The Gamesters of Triskelion, by Margaret
Armen**
Among the more thoughtful dramatic presentations of
slavery to be found anywhere, and no less exciting for its
thoughtfulness. There are the gladiator slaves to amuse
and delight you, along with the gambling instincts of
James T. Kirk. £50 £20

**No. 45  A Piece of Action, by David Harmon and Gene
Coon**
Anthropologists should stay awake nights worrying after
a glance at this episode. A crewman from the USS
Horizon leaves behind a book about Earth's Chicago
gangs of the 1930s. The cast includes Anthony Caruso as

the Boss, Victor Tayback, Lee Delano, John Harman,
Sheldon Collins and Steve Arnold.                          £50     £20

**No. 46  The Immunity Syndrome, by Robert Sabaroff**
Spock is the first to know it when he senses the death of
some 400 Vulcans aboard the ship Intrepid. On orders to
investigate, the Enterprise finds the sister ship has
disappeared. Excitement and mystery follow. Excellent
visuals (credit to Frank Van Der Veer) and an
authoritative success.                                     £50     £20

**No. 47  A Private Little War, by Gene Roddenberry**
On a peaceful planet Kirk has helped to explore some
years before, the crew finds that gunpowder has now
made an appearance. When a Klingon craft is discovered
in the area, the logical presumption is that they have
armed one group at the expense of another. Kirk is soon
on the scene to see to matters. This is a familiar tale of war
and a good episode.                                        £50     £20

**No. 48  Return to Tomorrow, by John Knightsbridge**
Three beings are in search of human bodies to house them
while they construct androids for themselves. Their
search seems ended when they find Kirk, Spock and a
doctor aboard the Enterprise, for which risk is its
business. Special guest star Diana Muldaur has a good
part in this story.                                        £50     £20

**No. 49  Patterns of Force, by John Meredyth Lucas**
A really scaring tale in which the Enterprise crew
discover a culture based on Nazism complete with
swastikas and the SS. The Jews of the piece are the
neighbouring (minority) Zeons. Kirk and McCoy are
clever in disguise as the SS to foil a self-styled Hitler.    £50     £20

**No. 50  By Any Other Name, by Dorothy Fontana and
Jerome Bixby**
Answering a distress call from a supposedly uninhabited
planet, the Enterprise and its crew are overpowered by
aliens who have taken on a human form. Rojan, played
by the star Warren Stevens, and his followers
commandeer the Enterprise for their intergalactic return
to their world, or so they think.                          £50     £20

**No. 51  The Omega Glory, by Gene Roddenberry**
Perhaps a story of greater impact than when first shown.
The moral is simple: documents of freedom and liberty
without constant practice of their stated virtues. Guest
stars are Morgan Woodward, Roy Jensen, Irene Kelly,
David L. Ross, Ed McReady, Lloyd Kino and Morgan
Farley. A splendid cast.                                   £50     £20

**No. 52  The Ultimate Computer, by D.C. Fontana** (from a story by Lawrence N. Wolfe)
The underlying dream of the technological era (a computer that can think) is a reality in this episode — a constant theme of modern science fiction, going back at least as far as Mary Shelley's *Frankenstein*. Partly to do with No. 31. William Marshall plays the part of Daystrom here.                                                                £50     £20

**No. 53  Bread and Circuses, by Gene Roddenberry and Gene Coon** (from a story by John Kneubuhi)
A TV game show called Name the Winner where gladiators fight it out Roman style? Well, it hasn't happened yet, but you never know. The defrocked Captain of the Starship Beagle (Merik played by William Smithers) has this bright idea to entertain the subjects of a dictatorship modelled after that of Imperial Rome.       £50     £20

**No. 54  Assignment Earth, by Gene Roddenberry** (from a story by Gene Roddenberry and Art Wallace)
This episode was a pilot for a Roddenberry series that never got off the ground. It is the story of a human named Gary Steven, played by Robert Lansing, who has been raised by aliens and sent to Earth to prevent a nuclear disaster. The episode includes footage of the Saturn V launching from Cape Kennedy. Guest stars include Terri Garr, Don Keeferz and Isis the cat.                        £50     £20

**No. 55  Spock's Brain, by Lee Cronin**
A beautiful apparition emerges on the Enterprise, and when she disappears, so has Spock's brain. McCoy manages to keep Spock's body alive and walking mechanically, but only temporarily. Guest stars are Marj Dusay, Sheila Leighton and James Daris.                £50     £20

**No. 56  The Enterprise Incident, by Dorothy Fontana**
Kirk makes a seemingly foolish error in crossing the Romulan barriers, only to have the Enterprise captured. However, it is a feather in the cap of the Romulan Commander (a great performance from Joanne Linville), especially as Spock is seriously thinking of joining her. A good plot with the new 29in. miniatures designed by Matt Jefferies for the Klingons.                         £50     £20

**No. 57  The Paradise Syndrome, by Margaret Armen**
An asteroid rushes towards an unfamiliar planet. Beaming down, the crew find a civilisation indistinguishable from that of the more advanced American Indian tribes. Kirk disappears, but William

Shatner seems to enjoy the release and turns in what I believe to be the performance of his lifetime. See it yourself!      £50    £20

### No. 58 And the Children Shall Lead, by Edward J. Lasko

Beaming down to a planet in distress, the crew finds that a wave of mass suicide has decimated the population. Five children have survived, playing some very odd games right out of *Lord of the Flies*. The children were responsible for the deaths of their parents, guided by the Gorgon. An unusual but great episode.      £50    £20

### No. 59 Is There No Truth, No Beauty?, by Jean Lisette Aroeste

A subtle, touching story of the Medusan Ambassador who comes aboard with his beautiful assistant. Diana Muldaur guest stars as the interpreter Miranda.      £50    £20

### No. 60 Spectre of the Gun, by Lee Cronin

At first glance the old 'time machine' theme again, this episode seems only to have transported Kirk, Spock etc. to a re-enactment of the film Gunfight at the OK Corral. The crew of the Enterprise are cast as the Clanton gang, trying to get a quality as in another film High Noon. Interesting.      £50    £20

### No. 61 Day of the Dove, by Jerome Bixby

In this story of violent trends, first the Enterprise crew are trying to kill each other, and then a dangerous game is set in motion. Guest stars are Michael Ansara and Susan Howard. One for the episodes of greats.      £50    £20

### No. 62 For the World is Hollow, and I have Touched the Sky, by Rik Vollarts

McCoy's prognosis of his disease is that he has only a year to live. Kate Woodville plays the High Priestess. Do not miss this exciting story.      £50    £20

### No. 63 The Tholian Web, by Judy Burns and Chet Richards

Mutiny is hardly a regular occurrence in the Starfleet, but that's what appears to have happened aboard the USS Defiance and four of the Enterprise crew beam aboard to find that everyone on the ship has murdered everyone else. Full credit to Frank Van Der Veer.      £50    £20

### No. 64  Plato's Stepchildren, by Meyer Dolinksy
The ideals of Ancient Greece seem to have had less effect on this civilisation than fashions. Guest starring is Michael Dunn (of Broadway theatre roles, including the Ballad of the Sad Café, and films such as Ship of Fools), a great actor and an even greater human being, who was short only in stature. He is a credit to this tale.      £50    £20

### No. 65  Wink of an Eye, by Lee Cronin
An interesting and well-handled story of survival, where the only clues are the story of a dead civilisation and a strange buzzing sound in the ears of the Enterprise crew. Could there be insects aboard? Search for this episode with its excellent slow-motion photography.      £50    £20

### No. 66  The Empath, by Joyce Muskat
Visiting a planet whose sun is about to become a nova, Kirk, Spock and McCoy find no trace of the previous people. Instead they find themselves beamed below the surface of the planet where a beautiful, though mute, woman awaits them. Kathryn Hays plays a great part, capable of empathetic cures, though at great cost to her own system.      £50    £20

### No. 67  Elaan of Troyius, by John M. Lucas
The fabulously beautiful Elaan, who is practically a god to her people, has become a pawn to her power, and she must marry the Chief of Troyius, her people's greatest enemy. France Nuyen is great as Elaan, as are the other players, Jay Robinson, Tony Young, Lee Duncan and Victor Brandt.      £50    £20

### No. 68  Whom Gods Destroy, by Lee Erwin
A federation of that insane asylum erupts in violent rebellion, while Kirk and Spock are drawn into the battle. The leader of the revolt, Garth, has an advantage over the Enterprise crew. Steve Innat plays Garth, with other guests like Keye Luke (of Charlie Chan fame as the number one son many years ago) and Yvonne Craig. Worth collecting.      £50    £20

### No. 69  Let That be Your Last Battlefield, by Oliver Crawford (from a story by Lee Cronin)
The tired old defiant clichés are played out in this indictment of racial (or, more specifically, simple colour) prejudice. Two aliens are split right down the middle, half black shoe polish, half white. Frank Gorshin plays the baddie in a James Cagney role, with Lou Antonio as the oppressed figure.      £50    £20

**No. 70 The Mark of Gideon, by G.F. Slavin and S. Adams**
Kirk is to be the first outsider to visit the Garden-of-Eden planet of Gideon, but the transporter sends him instead to a duplicate Enterprise, where there is no one else on board except for Odona, played by Sharon Acker. Other guest stars include David Hurst, Gene Kynarski and Richard Derr. £50 £20

**No. 71 That Which Survives, by John M. Lucas** (from a story by Dorothy Fontana)
A fairly extreme version of all those lethal lovelies of whom Raymond Chandler and Ian Fleming so often wrote. Lovelies are centre stage in this episode. Lee Meriwether plays Losira and is joined by Arthur Batanides and Naomi Pollack. £50 £20

**No. 72 The Lights of Zetar, by J. Tarcher and Shari Lewis**
Scott is sweet on a new female crew member of the Enterprise, but what at first seems to be a matter of her 'getting her space legs' becomes considerably more serious. Lieutenant Mira Romaine, played by Jan Shutan, is oddly affected by a strange phenomenon (visualised as a cluster of lights) trailing the ship. £50 £20

**No. 73 Requiem for Methuselah, by Jerome Bixby**
In the book of Genesis, 'all the days of Methuselah were nine hundred sixty and nine years', then, in the Bible's succinct way of handling these matters, he died. Author Bixby suggests he didn't die at all, but rather went on, cropping up as one of the great historical figures as Flint, played by James Daly, who lives in seclusion from the universe with a single companion, Reena, played by Louise Store, who loves to receive visitors, especially Kirk. £50 £20

**No. 74 This Way to Eden, by Arthur Heinemann** (from a story by Arthur Heinemann and Michael Richards)
Taking into consideration the time of this episode's original screening – just a month after President Nixon's first inauguration – the political implications of this tale of mad hippies from the 23rd century may be somewhat hard to take. Charles Napier stars, while the craft Aurora can be recognised as the Tholian Spider Ship from No. 63. £50 £20

**No. 75  The Cloud Minders, by Margaret Armen** (from
a story by David Gerrold and Oliver Crawford)
The story is not dissimilar to that of Fritz Lang's
Metropolis: the subterranean working-class slaves
provide the luxury required by the upper classes. Guest
star Fred Williamson is very good indeed.                    £50      £20

**No. 76  The Savage Curtain, by Arthur Heinemann**
(from a story by Gene Roddenberry)
The stories in which Roddenberry was personally
involved tend to be heavy with ideas. This does not
necessarily mean 'heaviness', for a strong story can
always carry a considerable weight of thought. Here Kirk
and Spock join forces with Abe Lincoln, played by Lee
Bergere, and the father of the Vulcan philosophy in a
battle against four men of evil, especially Genghis Khan,
played by Nathan Jung.                                       £50      £20

**No. 77  All Our Yesterdays, by Jean Lisette Aroeste**
A fairly 'time portal's' story takes on more than standard
interest as Spock evidences some mild romantic interest,
while Kirk is almost burned as a witch. On a doomed
planet the locals have disappeared. Watch this episode
and find the answers.                                        £50      £20

**No. 78  Turnabout Intruder, by Arthur Singer** (from a
story by Gene Roddenberry)
Talk about twist of a plot – this one centres around a mind
switch between Kirk and the scientist Dr Janice Lester,
played by Sandra Smith. Shatner manages to suggest a
female mind in his male body, while the woman scientist
desperately wants to become a man.                           £50      £20

## THE ANIMATED STAR TREK

The disembodied voices of the Enterprise crew, except Walter Keonig who
played Ensign Chekov, returned to the airwaves in 1972. As a network
presentation, the animated Star Trek lasted for 2 seasons, with 16 episodes in the
first year, and 6 in the second. In October 1976 Paramount Television announced
that it had syndicated the animated version to at least 25 markets, which was a
very impressive figure.

I would like to pay particular tribute to that very busy lady, Dorothy Fontana.
I am very grateful to the following prime producers, Lou Scheimer, Norm
Prescott, Hal Sutherland, William Shatner, Leonard Nimoy, DeForest Kelley,
James (Scottie) Doohan, Nichette Nichols, George Takei and Majel Barrett, and
all the wonderful technicians who are too numerous to name, together with all
the kind people whom I met in their many groups.

**No. 1  Beyond the Farthest Star, by Samuel A. Peeples**
Drawn to an alien starship billions of years old, the
Enterprise does not know that inside the ship awaits a
malevolent being seeking a new 'body'. This is a chilling
tale by the same writer who wrote the second pilot,
'Where No Man has Gone Before' (No. 3 of Episodes
1–78).                                                        £50        £20

**No. 2  Yesteryear, by Dorothy Fontana**
Sounding like something from the famous 'Matchbox
Old-Time Car Collection', in this story Spock must return
via the time portal (from 'City on the Edge of Forever',
No. 27 of Episodes 1–78) to Vulcan of 30 years past to
correct a distorted time line or he will die. Mark Leonard
is a guest star involved in Spock's boyhood.                 £50        £20

**No. 3  More Tribbles, More Troubles, by David Gerrold**
In this sequel to 'The Trouble with Tribbles' (No. 43 of
Episodes 1–78), the Enterprise rescues a small scout ship
from a Klingon battle cruiser. The crew find that they
have saved Cyrano Jones, while Stanley Adams guests as
his voice.                                                    £50        £20

**No. 4  The Survivor, by James Scherer**
This is a beauty and the beast story in which a woman
security officer believes that the survivor of a damaged
trader ship just beamed aboard the Enterprise is her
long-lost fiancé. However, things differ.                    £50        £20

**No. 5  Mudd's Passion, by Stephen Kandel**
In Kandel's third outing with his lovable galactic con
man, Harcourt Fenton Mudd, played by Roger C.
Carmel, is now peddling a love potion that really works.     £50        £20

**No. 6  The Magicks of Mega-Tu, by Larry Brody**
The Enterprise is on a mission to chart the core of the
Galaxy and is pulled into a different 'Time-space'
continuum. This has guest star Ed Bishop, who played
Commander Straker in UFO.                                     £50        £20

**No. 7  The Lorelei Signal, by Margaret Armen**
The Enterprise is lured to a strange golden planet by an
irresistible signal. The men all seem to be enchanted, and
the women on the ship, led by Lieutenant Uhura, must
take over in order to save it.                                £50        £20

**No. 8  One of Our Planets is Missing, by Marc Daniels**
Daniels directed 14 live action episodes and is considered
one of television's best directors. This story deals with a
strange matter cloud of energy which is moving into the
Galaxy. Can Kirk save the situation and possibly the
world?                                                          £50      £20

**No. 9  The Infinite Vulcan, by Walter Koenig**
While conducting a plant and planetary exhibition, a
landing party discovers a scientist who is seeking a
perfect being from which he can build an army in order
to maintain peace in the Galaxy. The author is the same
Walter Koenig who played the part of Ensign Chekov in
the live action episodes.                                      £50      £20

**No. 10  Time Trap, by Joyce Perry**
The Enterprise and a Klingon battle cruiser are both
drawn into the Delta Triangle and find out that it is a time
trap from which they may never escape.                        £50      £20

**No. 11  The Slaver Weapon, by David Harmon**
Spock, Sulu and Uhura are using a shuttlecraft to
transport a Stasis Box, a rare and valuable relic of the
Slaver Empire, when they detect another such box and go
to investigate. This fine story won more than just praise,
it won awards.                                                £50      £20

**No. 12  The Jihad, by Stephen Kandel**
In Kandel's second script for the animated series, Kirk
and Spock are chosen to join a top secret team of aliens to
accomplish a vital mission. This was one of the best
animated episodes of all time.                               £50      £20

**No. 13  The Ambergris Element, by Margaret Armen**
Like Kandel, Armen wrote two episodes for the animated
series. In this story, an Enterprise party takes an
aquashuttle (a sort of hovercraft) down to a water planet
and is attacked by a large sea creature, possibly the Loch
Ness monster on holiday from Scotland visiting its
American cousins. The shuttle is destroyed and Kirk and
Spock get lost.                                               £50      £20

**No. 14  Once Upon a Planet, by Lynne L. Janson and
Chuck Menville**
In this unsatisfactory adaptation of Theodore Sturgeon's
'Shore Leave' (No. 14 of Episodes 1–78) the recreation
planet has now become a very dangerous spot to be in.
This poor story does however offer you the wonderful
voice of Nichette Nichols.                                    £50      £20

**No. 15  The Terratin Incident, by Paul Schneider**
After receiving a mysterious signal, the Enterprise follows the strange transmission back to its source. Soon after this, members of the crew begin to shrink. Schneider also wrote the highly acclaimed and award winning 'The Balance of Terror' and 'The Squire of Gothos' (Nos. 13 and 16 in Episodes 1–78).      £50     £20

**No. 16  The Eye of the Beholder, by David Harmon**
Investigating the disappearance of a science team from another ship, Kirk, Spock and McCoy beam down to the surface of a planet and find themselves in a 'Zoo'. The Zookeepers look like giant slugs, but have IQs in their thousands. Delightful with cute aliens.      £50     £20

**No. 17  The Pirates of Orion, by Howard Weinstein**
Spock has contracted a disease fatal to Vulcans, and his life depends upon the delivery of a rare drug.      £50     £20

**No. 18  Albatross, by Dario Finelli**
During a stop at the remote planet of Damia, the Enterprise crew is stunned when McCoy is arrested for the slaughter of hundreds of Damians, 1900 years before. All in all, quite interesting.      £50     £20

**No. 19  Bem, by David Gerrold**
Bem is the representative of an alien race considering membership of the Federation. The Enterprise and her crew are among a group being tested by Bem's people as to the worthiness of the Federation. They may fail. A good story is flawed towards the end of the plot by the character who bears a disgracefully similar resemblance to the Companion from 'Metamorphosis' (No. 37 in Episodes 1–78).      £50     £20

**No. 20  Practical Joker, by Chuck Menville**
The main computer of the Enterprise wreaks havoc on the crew when electronic particles invade its circuits. The hilarious farce is highlighted by our first glimpse of the ship's recreation room, where the illusion of being anywhere in any world is simply a matter of programming the room's controls. Roddenberry first created this concept when putting together the live action series, but budget considerations intervened.      £50     £20

**No. 21  How Sharper than a Serpent's Tooth, by Russell Bates and David Wise**
The Enterprise encounters a vessel that soon cripples it. Kulkulkan, a god who actually existed according to legend, is the pilot of the enemy ship and demands recognition. A very good and intriguing story.      £50     £20

**No. 22  Counter-Clock Incident, by John Culver**
An alien ship pulls the Enterprise and her crew
(accompanied by Commodore Robert April, the first
captain of the Enterprise, and his wife) into a nova. There,
in a reverse universe, black stars shine in the whiteness
of space, and people who are born old then die young
(quite a unique plot). The trapped crew begin to turn into
children, losing their knowledge and space skills. The
end of the episode offers a strong moral. April and his
wife, who have decreased in age about 40 years, and are
now 25, decide that it is better to stay old and cherish fond
memories of the past than to become young again. Very
well done and one of the best ever in the annals of space
stories.                                                          £50      £20

## THE SECOND GENERATION

The following highly collectable videos were made by Paramount exclusively
for the F.W. Woolworth stores. In order to get full collector value, the videos
must be unused and boxed in their original sealed wrapping.

**No. VHR 4101  Star Trek TV Movie 1, 'Encounter at
Farpoint'**
Price £12.95. Issued 1994. Deleted April 1995.                    £20       £5

**No. VHR 4102  Star Trek TV Movie 2, 'The Best of Both
Worlds'**
Price £12.95. Issued 1994. Deleted April 1995.                    £20       £5

**No. VHR 4103  Star Trek TV Movie 3, 'Redemption'**
Price £12.95. Issued 1994. Deleted April 1995.                    £20       £5

**No. VHR 4104  Star Trek TV Movie 4, 'Unification'**
Price £12.95. Issued February 1995. Deleted April 1995.           £20       £5

**No. VHR 4105  Star Trek TV Movie 5, 'Time's Arrow'**
Price £12.95. Issued March 1995. Deleted April 1995.              £20       £5

**Star Trek Generations**
Video guide to the above five TV movies given free with
tokens collected from newspapers. Approx. 21 mins.                £20       £5

**Woolworth Leaflet**
This advertising leaflet contains five £2 tokens which
could be used to purchase the videos. Price is for mint
and unused tokens.                                                £10       £2

## STAR TREK SERIES

Short-story adaptations by James Blish of the TV episodes. Originally published by Bantam Books at US$1.25 each. Also sold in boxed sets of six copies (value of a set in mint condition: £200).

| | MC | GC | FC |
|---|---|---|---|
| **Star Trek 1**<br>Adaptations of 'Charlie's Law', 'Dagger of the Mind', 'The Unreal McCoy', 'Balance of Terror', 'The Naked Time', 'Miri' and 'The Conscience of the King'. | £30 | £15 | £10 |
| **Star Trek 2**<br>Adaptations of 'Arena', 'A Taste of Armageddon', 'Tomorrow is Yesterday', 'Errand of Mercy', 'Court Martial', 'Operation Annihilate', 'The City on the Edge of Forever' and 'Space Speed'. | £30 | £15 | £10 |
| **Star Trek 3**<br>Adaptations of 'The Trouble with Tribbles', 'The Last Gunfight', 'The Doomsday Machine', 'Assignment Earth', 'Mirror Mirror', 'Friday's Child' and 'Amok Time'. | £30 | £20 | £10 |
| **Star Trek 4**<br>Adaptations of 'All Our Yesterdays', 'The Devil in the Dark', 'Journey to Babel', 'The Menagerie', 'The Enterprise Incident' and 'A Piece of the Action'. | £30 | £20 | £10 |
| **Star Trek 5**<br>Adaptations of 'Whom the Gods Destroy', 'The Tholian Web', 'Let that be Your Last Battlefield', 'Requiem for Methuselah' and 'This Way to Eden'. | £30 | £20 | £10 |
| **Star Trek 6**<br>Adaptations of 'The Savage Curtain', 'The Lights of Zetar', 'The Apple', 'By Any Other Name', 'The Cloud Minders' and 'The Mark of Gideon'. It won the International Hugo Award for best dramatic presentation of the year and was also voted the best TV script of the year. | | | |
| Unsigned | £30 | £20 | £10 |
| Signed | £1000 | — | — |
| **Star Trek 7**<br>Adaptations of 'Who Mourns for Adonis?', 'The Changeling', 'The Paradise Syndrome', 'Metamorphosis', 'The Deadly Years' and 'Elaan of Troylus'. | £30 | £20 | £10 |

**Star Trek 8**
Adaptations of 'Spock's Brain', 'The Enemy Within', 'Catspaw', 'Where No Man has Gone Before', 'Wolf in the Fold', 'For the World is Hollow' and 'I have Touched the Sky'. £30 £20 £10

**Star Trek 9**
Adaptations of 'Return to Tomorrow', 'The Ultimate Computer', 'That Which Survives', 'Obsession', 'The Return of the Archons' and 'The Immunity Syndrome'. £30 £20 £10

**Star Trek 10**
Adaptations of 'The Alternative Factor', 'The Empath', 'The Galileo Seven', 'Is There in Truth No Beauty?', 'A Private Little War' and 'The Omega Glory'. £30 £20 £10

**Star Trek 11**
Adaptations of 'What are Little Girls Made of?', 'The Squire of Gothos', 'The Wink of an Eye', 'Bread and Circuses', 'The Day of the Dove' and 'Plato's Stepchildren'. £30 £20 £10

## OTHER PUBLICATIONS

**Star Trek Blueprints, by Franz J. Schnaubelt**
Gives Starship Blueprint designs in detail as well as exterior and interior views. Sick bay, bridge, Spock's science lab and every level of the Enterprise in exact scale. Price US$1.25. Published 1975 by Ballantine Books. £50 £20 £10

**Star Trek Book and Record**
Book and record set. Published by Peter Pan Industries. Good investment. £100 £50 £10

**Grosset & Dunlap's Star Trek Catalogue**
Features many important details, addresses and pictures of the Star Trek cast. Published 1979 by Grosset & Dunlap. The price came down to only 50p in the final stages of sale, but the catalogue was a wise investment for any person who could see the future in space collectables. Copies were often destroyed, which means a high value for collectors today. Any signature will boost the value of a copy.

| | MC | GC | FC |
|---|---|---|---|
| Unsigned | £100 | £50 | £20 |
| Signed | £500 | £300 | £20 |

**Star Trek Concordance, by Bjo Trimble**
Complete encyclopedia and cross-reference of people, places and things in Trekdom, with drawings and photos. Price US$5.95. Published by Ballantine Books. £50 £30 £15

**Letters to Star Trek, edited by Susan Sackett**
Introduction by Gene Roddenberry. The best of the tons
of letters sent to Star Trek staff and stars, and the replies.
Price US$1.95. Published 1977 by Ballantine Books.

| | MC | GC | FC |
|---|---|---|---|
| | £100 | £30 | £10 |

**Star Trek Lives, by Jacqueline Lichenberg**
Co-writers: Sondra Marshak and Joan Winston. In-depth
examination of the creators, crew, fans, conventions and
stars. Price US$1.95. Published 1975 by Bantam Books.

| | £100 | £30 | £15 |
|---|---|---|---|

**Star Trek Log, One through Nine, by Alan Dean Foster**
Price US$1.50 for each book. Published by Ballantine
Books.

| | MC | GC | FC |
|---|---|---|---|
| Single copy | £30 | £20 | £10 |
| Set of nine | £300 | — | — |

**Star Trek Log, One to Four**
Series containing adaptations of the TV animated series
in story form. Published by Ballantine Books.

| | MC | GC | FC |
|---|---|---|---|
| Single copy | £50 | £25 | £10 |
| Set of four | £200 | — | — |

**The Making of Star Trek, by Stephen E. Whitfield and
Gene Roddenberry**
Everything anyone could wish to know about the series:
How, when and where Star Trek was born, grew and was
born again. Price US$1.95. Published by Ballantine Books.

| | £50 | £20 | £10 |
|---|---|---|---|

**Star Trek: The New Voyages, by Myrna Culbreath and
Sondra Marshak**
New, original adventures and very exciting stories. Price
US$1.75. Published 1976 by Bantam Books.

| | £30 | £20 | £10 |
|---|---|---|---|

**The Star Trek Reader, by James Blish**
A very good book with adaptations of 21 TV scripts in a
hardback edition. Price US$8.95. Published 1976 by E.P.
Dutton.

| | £500 | £100 | £50 |
|---|---|---|---|

**Star Fleet Technical Manual, designed by Franz Joseph**
Complete and wonderful guide to the Enterprise and the
United Federation of Planets. Includes the Articles of
Federation, all the insignia of the Armed Forces of the
Federation, detailed descriptions of the Starships in the
Fleet etc. Many technical drawings accompany the text.
Price US$6.95. Published 1975 by Ballantine Books.

| | £200 | £75 | £35 |
|---|---|---|---|

**Trek or Treat, by Eleanor Ehrhardt and Terry Flanagan**
A lovely picture book of the Trek stars with space balloon
captions that are as amusing as Blooper reels. Price
US$2.95. Published 1977 by Ballantine Books.

| | £50 | £20 | £10 |
|---|---|---|---|

**The World of Star Trek, by David Gerrold**
An absolutely fascinating inside view of the Trek saga, its
past, present and future. It makes you feel like part of the
crew. Price US$1.95. Published 1973 by Ballantine Books.   £50   £25   £15

**I am not Spock, by Leonard Nimoy**
Where Spock ends and the real Leonard Nimoy begins.
A charming autobiography. Price US$4.95. Published
1976 by Celestial Arts.   £50   £20   £10

**Spock-Messiah, by Theodore R. Coggswell and Charles
A. Spano Jr**
A real chiller-thriller adventure book. Price US$1.75.
Published 1976 by Bantam Books.   £50   £20   £10

**Spock must Die, by James Blish**
The master science-fiction writer's original novel based
on the Spockman. Price US$1.25. Published 1979 by
Bantam Books.   £50   £20   £10

**Will I Think of You?, by Leonard Nimoy**
Nimoy's deeply emotional love poem. His word pictures
are as beautiful as his photography. A rare and much
sought after book, especially copies signed by the author.
Price US$3.95. Published by Celestial Arts.

| | MC | GC | FC |
|---|---|---|---|
| Unsigned | £50 | £30 | £20 |
| Signed | £2000 | — | — |

**You and I, by Leonard Nimoy**
By the star who loves to write about the series he loves.
Superb photographs and poems beautifully presented.
Price US$3.95. Published by Celestial Arts.

| | MC | GC | FC |
|---|---|---|---|
| Unsigned | £100 | £50 | £20 |
| Signed | £2000 | — | — |

# COMICS AND MAGAZINES

Nobody in the history of the press could have predicted that a weekly children's comic could ever have become a collectable, let alone be worth a staggering £25000. However, *Action Comic*, No. 1, June 1939 is indeed worth that sum, and furthermore it started the adventures of possibly the greatest hero of all time, Superman, the brainchild of Jerry Siegal and Joe Shuster. My thanks go to all the kind collectors who assisted me with the comics and magazines.

| COMIC/MAGAZINE | MC | GC | FC |
|---|---|---|---|
| **Action Comic, No. 1, June 1939**<br>This Superman hero figure had been tried in various ways to get publishers interested but had failed. This particular hero was the brainchild of Jerry Siegal and Joe Shuster. Although any early and later issues telling the tales of this most famous of all spacemen from this era are worth collecting, it is the very first issue that is so valuable. | £25000 | — | — |
| **Action Comic, No. 285, February 1961**<br>At last Superman introduced Supergirl to the world and his millions of fans. Colourful cover fully shows a banner with the words 'Welcome Supergirl' in pink lettering and underneath the figures of Superman and Supergirl in flight. Published by National Periodical Publications. | £250 | £50 | £10 |
| **Adult Tales of Terror, No. 1, 1955**<br>Horror and suspense illustrated, introducing Picto-Fiction. Covers by Reed Crandall. Published by IC Publishing. | £50 | £20 | £5 |
| **Adventure Comic, No. 75, 1942**<br>With Starman, the Sandman and Sandy. Became one of the top selling weeklies in the history of comics. | £100 | £20 | £5 |
| **Adventures into the Unknown, by Richard Hughes, 1950s** | £100 | £30 | £10 |
| **Adventures into the Unknown, No. 117, June/July 1960**<br>With Spencer Special. Published by Bert Publishing. | £100 | £30 | £5 |

**Adventures into the Unknown Comic, No. 1, 1951/53**
With space demons from the outer planets, this was
America's first supernatural comic. Published by B & I
Publishing.                                                £250    £50    £10

**The Adventures of the Fly, No. 1, August 1959**
First of the Archie adventure series with a colourful cover
from Simon and Kirby, 'The Fly Meets Spider Spry'. A
real winner. The series ended in 1964. Published by Radio
Comics.                                                    £250    £50    £10

**All-American Comic, No. 1, 1939**
The adventures of Ultra Man set in the year 2239.
Artwork by Jon L. Blummer, but issues are rare,
especially Nos. 1 and 8.                                   £100    £40    £10

**All-Star Comic, No. 1, Summer 1940**
A colourful front cover with four action scenes made it
an instant winner. The designer Sheldon Mayer began to
chronicle the adventures of the Justice Society of
America.                                                   £250    £50    £10

**All-Winners Comic, No. 1, Summer 1941**
This favourite comic of course contained the popular
Captain America and Bucky, the Human Torch and Toro,
the Angel and Black Marvel, as well as the Sub-Mariner
– all great heroes in their own right. Published by Comic
Magazine Corp.                                             £200    £30    £5

**Amazing Fantasy, No. 1, 1962**                           £100    £30    £5

**Amazing Fantasy, No. 15, 1962**
Introduced Spiderman to the world. Artwork by Jack
Kirby. Price US12c. Published by Atlas Magazines.          £100    £30    £5

**The Amazing Spiderman, No. 1, 1963**
Although merely a teenager, Spiderman had to deal with
some evil foes, the most intriguing and intimidating of
whom was Doctor Doom, the mad genius. Published by
Non-Pareil Publishing.                                     £1000   £250   £50

**The American Flag, No. 1, 1983**
Became one of the best selling comics of the 1980s.
Published by First Comics and Howard Chaykin.              £20     £10    £5

**America's Best Comic, No. 1, 1946**
With Doc Strange and the Black Terror. Designed mainly
by Alex Schomburg, one of the many overseas artists to
seek work in the US. Published by Visual Publications.     £200    £50    £10

**The Arrow, No. 1, 1940**
The next hero to come on to the scene after Superman.
Created by Paul Gustavson.          £200     £30     £10

**Avengers, No. 1, 1964**
One to add to the collector world with adventures galore
on Earth and in space. Published by Marvel Comics.          £200     £30     £10

**Avengers, No. 4, 1964, Captain America Lives Again**          £100     £20     £5

**Batman, No. 1, Spring 1950**
The second most important superhero was Batman, the
mystery avenger who patrolled Gotham City with his
young partner, Robin the Boy Wonder. The cover of the
first issue saw the two heroes flying through space and
was to last for many years. Published by DC Comics.          £5000     £200     £20

**Batman and the Joker's Utility Belt, No. 73, 1952**
Dick Sprang, who for more than 20 years drew the comic
book adventures of Batman, was hardly ever allowed to
sign his work. However, when Sprang drew the first
pictures of the Caped Crusader in 1943, it is believed that
he autographed a copy. Also I was in conversation with
a collector of these comics who said that he had seen a
signed copy of No. 73. A million pounds would not buy
such a copy if it existed, but you never know...          £500     £100     £30

**Batman, No. 160, December 1954**
Of the many conceptions of Superman that were drawn
over the 50-plus years, Wayne Boring's is the best work
assisted by Stan Kaye. This issue is one of the best
examples work of a quality investment.          £250     £50     £10

**Batman Goes to Camp, 1966**
A further collector's item. Art done by Carman Infantino
and Murphy Anderson. Published by National Periodical
Publications.          £200     £30     £10

**Batman, The Killing Joke, Special Issue, 1988**
The 1980s killed certain comic publishers stone dead,
while others prospered. Batman and Superman had both
been around for 50 years but with new writers, art men
and techniques, they began to reap in millions. DC
Comics and black and white limited editions did well.
Look out for these regardless of title or character since
they are like gold.          £50     £20     £5

**Black Cat Mystery Comic, No. 1, 1951/53**
Published by Harvey Features Syndicate at a time when
even Hollywood was interested in the comic world.          £100     £30     £5

|---|---|---|---|

**The Blue Beetle, No. 1, 1939**
With the exciting adventures of another famous fighter for justice. Published by Fox Publications.

|  | MC | GC | FC |
|---|---|---|---|
| The Blue Beetle, No. 1, 1939 | £250 | £50 | £10 |

**Blue Beetle Comic, 1948**
With females in chains, others to break the chains and more space creatures than ever before. Published by Fox Publications. Good investment series.

£200 — £50 — £10

**Blue Ribbon Comic, No. 1, 1940**
With famous characters, action mystery and thrills, featuring Wang-a-Tang the Wonder Dog drawn by Ed Smalle. Published by MLJ Magazines.

£100 — £30 — £5

**Bulletman Comic, No. 1, 1941**
Featuring Bulletgirl with 64 exciting pages, it appealed to both sexes, young and old alike. It was amazing just how many grown-up men bought and hid their comics with the women heroes. Published by Fawcett Publications.

£200 — £50 — £10

**Camelot 3000, Nos. 1–12, 1982**
Created by Mike Barr and Brian Bolland. Published by DC Comics. Set of 12.

£50 — — — —

**Captain America, No. 1, December 1945**
The ultimate patriot Captain America became one of the greatest comic heroes of all time. His adventures from the war years are both thrilling and memorable, especially his fight against the evil Hitler in No. 3. Published by Complete Story Corp.

| | MC | GC | FC |
|---|---|---|---|
| No. 1 | £2000 | £200 | £50 |
| No. 3 | £5000 | £500 | £100 |

**Captain Flash, No. 1, 1954**
Feature page by designer artist Mike Sekowsky. Published by Sterling Comics.

£200 — £50 — £10

**Captain Marvel Junior, No. 1, November 1942**
So successful was the Marvel name that a new part of the world brought Captain Marvel Junior. Talented Mac Raboy drew the cover. Published by Fawcett Publications.

£250 — £50 — £10

**Captain Midnight, No. 1, September 1942**
With success through the 1940s, things could only get better when Captain Marvel introduced his friend Captain Midnight, who turned out to be one of the great spacemen of all time, a star also of radio and films. Published by Fawcett Publications.

£250 — £50 — £10

**Captain Tootsie, No. 1, 1950**
Included 'Rocket to the Planet Venus' and 'The Secret
Legion'. Price US10c. Published by Toby Press.

| | MC | GC | FC |
|---|---|---|---|
| Captain Tootsie, No. 1, 1950 | £100 | £30 | £5 |

**Cerebus the Aardvark, No. 1, 1977**
A great commercial success by Dae Simms, who
produced it and also did the artwork.

£30 £10 £5

**Crack Comic, No. 1, Spring 1940**
Brought more famous characters like the Space Legion
and the Red Torpedo as well as Madam Fatal and the
Black Condor.

£500 £50 £10

**Crusader from Mars, 1951**  £100  £30  £5

**Daredevil Comic**
One of the greatest titles in comic history, Daredevil was
to become the world's leading man of action for millions.
Published by Lev Gleason Publications.
No. 1, 1941. Rare.  £1000  £100  £30
No. 2, 1941.  £500  £50  £20

**Daredevil Comic, 8 June 1960**
Here is a real treasure: the comic that printed in brown
letters 'Here Comes the Man without Fear', then in large
bold red and black letters 'Daredevil'. In this issue
Daredevil is versus the Stilt Man. Marvel Comics' clever
variations on this superhero during the 1960s seemed
endless. Daredevil may have been the most ingenious
creation of all; he was totally blind but aided in his career
by an Infallible Radar Senser. Cover for this issue by
Wally Wood. Price US12c. Top investment.

£200 £50 £10

**Daredevil Comic, No. 181, April 1982**
Many people objected to this issue which included a
degree of violence never before seen in any comic. It
contains the vicious fight between Bullseye and Elektra,
when he not only kills his opponent, but relishing every
blow, breaks her jaw, slits her throat and thrusts a knife
through her body. Therefore this issue is hard to come by.

£500 £100 £30

**The New Daredevil, No. 1, 1981**
Even more successful than the old series, it introduced
Elektra the Assassin, an evil female.

£40 £10 £5

**Daring Mystery Comic, No. 1, 1939**
Forerunner of eventual *Marvel Mystery Comic*, and then
just *Marvel*. Artwork by Martin Goodman. Rare and a
sound investment.

£2000 £200 £50

**The Dazzler, No. 1, 1981**
A first not only because the Female Beautiful Mutant worked as a disco singer, but because the publisher Marvel Comics made it the first direct sales comic. Expectations were high and dealers ordered some 400,000 copies. It was the first time that a major comic book had deliberately bypassed the traditional method of selling a comic on the news-stands. Here at last was recognition from the industry's largest and most successful publisher that direct sales shops had the clout to make or break individual titles, and it worked.    £200     £40     £5

**The Denver Post, 2 July 1933**
This weekly magazine had the comic-strip adventures of Buck Rogers. Rare.    £250     £50     £20

**Destination Moon, 1950**
With the famous Kris K-L-99, a space explorer, by Edward Hamilton and artist Howard Sherman.    £500     £100     £30

**Detective Comic, No. 27, 1939**
Contains the first appearance of Batman and Robin. Created by Bill Finger and artist Bob Kane. Worthwhile investment.    £250     £75     £25

**Detective Comic, No. 54, 1941**
Beautiful cover with Batman doing his acrobatic skills, probably designed and drawn by artist Jerry Robinson.    £250     £75     £25

**Dynamo Man, No. 1, 1966**
Cover by Wally Wood for the Thunder Agent. Published by Tower Comics.    £100     £30     £10

**The Eagle Comic, No. 1, July 1941**
Had everything that the space reader could wish for. Artwork by Jack Binder. Published by New Friday Publications.    £200     £40     £10

**Exciting Comic, No. 1, April 1940**
Offered a mixed bag, including the Space Rovers by Max Plaisted.    £250     £50     £10

**Famous Funnies, No. 3, 1950**
Forerunner of all US comics, promising over 100 characters for only a dime, during the Depression. It did not pick up until this third issue which featured Buck Rogers, first of the real spacemen. Rare.    £2000     £100     £25

**The Fantastic Four, No. 1, 1961**
Another successful venture into comic books by writer-editor Stan Lee. Artwork by Jack Kirby and Dick Ayers. Published by Marvel Comics.    £250     £30     £10

**The Fantastic Four, No. 51, June 1966**
The years brought success after success to this comic, and by the time of this issue sales had exceeded all expectations. Stan Lee and Jack Kirby had made the Orange Skinned Thing famous. — £50 £20 £5

**Feature Comic, No. 54, March 1942**
This issue introduced another space-flying hero, Dollman, whose success eventually brought him his own comic. Other issues are also worth collecting. Published by Comic Favourite Publishers. — £100 £20 £5

**The Flame Comic, No. 1, 1940**
Dishonest villains, beware of another hero! Illustrated by Lou Fine. Published by Fox Features Syndicate. — £250 £75 £25

**Flash Comic, No. 1, January 1940**
The character Flash was able to outrun and outstay most of his foes. Published by DC Comics. — £250 £50 £10

**The Forever People, by Jimmy Olsen, 1972**
Published by National Periodical Publications. — £30 £10 £5

**Funny Animals Comic, No. 1, December 1942**
So popular were the character and adventures of Captain Marvel that he was utilised by Fawcett Publications to launch a new title. — £250 £40 £10

**Green Lantern Comic, No. 1, Fall 1941**
The superhero Green Lantern proved to be very popular all over the world. Cover by Howard Purcell. Price US10c. Published by DC Comics. Rare. — £100 £30 £5

**Green Lantern Comic, No. 1, 1970**
Incorporating Green Arrow, this series became a great success in the 1970s. Published by National Periodical Publications. — £100 £30 £10

**Hawkman, Introducing Creature of a Thousand Shapes, 1961**
Revival of the 1940s character. Cover artwork by Joe Kubert. — £50 £10 £5

**Hawkman Books, No. 3, 1964**
Colourful cage with the title 'Birds in the Gilded Cage'. Cover art by Murphy Anderson. Published by National Periodical Publications. — £50 £10 £5

**Howard the Duck**
Popular and clever series about a duck trapped in a world
of his own. Created by Steve Gerber. Cover art by Frank
Brunner. Published by Marvel Comics.

| | MC | GC | FC |
|---|---|---|---|
| No. 1, 1975 | £50 | £20 | £5 |
| No. 2, 1975 | £30 | £10 | £3 |

| | | | |
|---|---|---|---|
| **I Sealed the Earth's Doom, 1954** | £100 | £30 | £10 |

**It Came from Outer Space, 1953**
One of the first 3-D comic books.

| | | | |
|---|---|---|---|
| | £250 | £50 | £20 |

**Jackpot Comic, No. 1, 1941**
Its stars strike a blow for freedom of the future world:
Steel Sterling, Black Hood, Mr Justice, S. Boyle and many
more. Published by MLJ Magazines.

| | | | |
|---|---|---|---|
| | £100 | £30 | £5 |

**Journey into Mystery Comic, No. 1, 1961**
Good comic for boys and girls of all ages. Published by
Marvel Comics.

| | | | |
|---|---|---|---|
| | £200 | £30 | £10 |

**Journey into Mystery Comic, No. 83, 1964**
In this issue the writer-editor Stan Lee found the
inspiration for Mighty Thor, who went on to become one
of the greatest stars published by Marvel Comics.

| | | | |
|---|---|---|---|
| | £200 | £30 | £10 |

| | | | |
|---|---|---|---|
| **Journey into Unknown Worlds, 1950** | £100 | £30 | £5 |

**Judge Dredd**
A British import into America. The Judge impressed
readers with his cynical humour. Splendid artwork of
Brian Bolland. Published by Tower Comics.

| | MC | GC | FC |
|---|---|---|---|
| No. 1, 1983 | £20 | £5 | £3 |
| No. 2, 1983 | £10 | £5 | £3 |

**Kamandi Comics, November 1972**
Featuring the Last Boy on Earth, this was a sensational
Jack Kirby blockbuster. The title borrowed ideas from
several sources, especially The Planet of the Apes. Cover
art by Kirby and Mike Royer. Published by National
Periodical Publications.

| | | | |
|---|---|---|---|
| | £50 | £20 | £5 |

| | | | |
|---|---|---|---|
| **Lars of Mars, 1950** | £100 | £30 | £5 |

**Life with Archie, No. 1, 1965**
This hero, inspired by Batman, went under the name of
Captain Pureheart seeking justice against evil on Earth
and in the Planet Zones. Published by Archie Comics.

| | | | |
|---|---|---|---|
| | £100 | £30 | £10 |

**Luke Cage, Hero for Hire, No. 1, 1972**
This successful character was made to happen by John
Romita, who also did the cover. Published by Marvel
Comics.

| | MC | GC | FC |
|---|---|---|---|
| | £30 | £10 | £5 |

**Man o' Mars, 1951** £100 £30 £5

**Marvel Comic, No. 1, 1940 (Special Edition Comics)**
Became famous with wild non-stop action, especially
with the Human Torch. £250 £50 £10

**Marvel Mystery Comic, No. 1, 1939**
Rare and a great investment. £5000 £500 £50

**Mary Marvel Comic, No. 1, December 1945**
This pretty heroine, as dedicated as her big brother
Captain Marvel, won her own magazine, which went into
great demand. Published by Fawcett Publications. £250 £50 £10

**Master Comic, No. 32, November 1942**
Another popular comic. With the adventures of Captain
Marvel Junior. Published by Fawcett Publications. £100 £30 £5

**Metal Men, No. 1, 1963**
The robot called Platinum attempts to save the life of her
creator on the cover of the first issue. Cover art by Ross
Ardu and Mike Esposito. Published by National
Periodical Publications. £200 £40 £10

**Metamorpho the Element Man, No. 1, 1964**
US12c. Published by National Periodical Publications.
Good investment. £250 £40 £10

**Mighty Mouse, by Phil Terry, 1953**
The very first of the 3-D comic books, rare with the first
pair of 3-D Mighty Mouse Space Goggles. Price US25c
with free goggles.

| | MC | GC | FC |
|---|---|---|---|
| With goggles | £500 | £50 | £20 |
| Without goggles | £300 | £30 | £10 |

**The Mighty Thor, No. 1, 1983**
This series became a top-selling title and gained the
Mighty Thor millions of fans. Published by Marvel
Comics. £100 £30 £5

**Mister Miracle, 1972**
Published by National Periodical Publications. £30 £10 £5

**Mr Mystery Comic, No. 1, 1952**
Colourful adventure comic with plenty to please the
many fans and collectors alike. With Wolverton. £200 £50 £10

**Moon Knight, No. 1, 1980**
In the tradition of the Shadow, the Moon Knight is there
to help all in danger from evil. Colourful cover art by Bill
Sienkiewicz.      £50    £20    £10

**More Fun Comics, No. 101, 1944**
Many people had asked about the childhood of
Superman, and the character of Superboy was created,
even though it caused court actions over rights etc.
Nevertheless this issue is a collector's gem.    £2000    £50    £10

**Mystery Comic, No. 8, 1940**
With a famous character called the Green Mask, well
advanced as a future world hero or villain. Published by
Fox Publications.      £100    £20    £5

**Mystery in Space, No. 1, 1954**
This great series of penal science fiction was an instant
international success. Published by DC Comics.    £300    £50    £20

**Mystery in Space, No. 18, 1956**
With its story of 9 prisoners from 9 planets, Chain Gang
of Space.      £200    £30    £10

**New Gods, No. 1, 1971**
Good selling series. Published by National Periodical
Publications.      £50    £20    £5

**The New Gods, by Jimmy Olsen, 1972**      £30    £10    £5

**New Teen Titans, No. 1, 1981**
Solid success. Cover and inside art by George Perez, who
was also responsible for the title. Published by DC
Comics.      £50    £10    £5

**Nexus Comic, No. 1, 1988**
Mike Barron and Steve Rudes smartly produced comic
which became one of the better independent superhero
titles of the 1980s. Published by First Comics.    £20    £10    £5

**Nick Fury Comic, No. 1, 1968**
The artwork for this first issue was coloured by Joe
Sinnott. Published by Olympia Group.    £100    £30    £5

**Nickel Comic, Nos. 1–8, 1940**
This comic only lasted eight issues but is most certainly
worth collecting. Covers by Jack Binder. Published by
Fawcett Publications.
Single issue      £100    £20    £5
Set of eight      £1000    —    —

**Noman the Thunder Agent, No. 1, 1966**
Cover coloured by Wally Wood. Published by Tower
Comics. £100 £30 £10

**Out of This World, 1956**
Gripping comic and collector's gem. Published by
Charlton. £100 £30 £5

**The Peacemaker, No. 1, 1967**
Well-received series with cover artists like Pat Boyette.
Published by Charlton Comics. No collector should miss
this series. £200 £40 £10

**Pep Comic, No. 1, 1941**
With the Shield and Dusty team, and the Hangman by
Irv Norvick, readers had the men flying through space in
all directions. Published by MLJ Magazines. £100 £30 £5

**Planet Comic, No. 1, 1946**
Attractive women were now a great part of the comic
book world, but they had to fight and be very successful.
This comic proved that monsters with one eye and purple
beaks had an eye for females, and many women were
easy prey. Published by Fiction House. £250 £50 £10

**Plastic Man Comic, No. 1, 1944**
By artist Jack Cole. Published by Vital Publications. £100 £30 £5

**Popular Comics, 1934–6**
This US series contained reprints from Famous Funnies.
Any title which included the Flash, Wonder Woman, the
Atom or Green Flash will bring the following prices. £100 20 £5

**Prize Comic, No. 1, 1940**
This magazine was to give readers the Green Lama and
the Black Owl, a real superhero flying through time and
space. Published by Feature Publications. £100 £30 £5

**Prize Comic, No. 1, 1942 (Second Series)**
Another winner which had the brave antics of boy
wonders Yank and Doodle. Published by Fawcett
Publications. Good investments. £100 £20 £5

**The Punisher, No. 1, 1987**
Successful hard-hitting character. Published by Marvel
Comics. £20 £10 £5

**Real Fact Comic, mid-1940s**
Several copies of the non-fictional stories were signed by
the famous artist Dick Sprang, and are rare as they were
soon snapped up. One signed copy was sold on a golf
course in Palm Springs for US$75000! Therefore I cannot
put a price on the signed copies, but unsigned copies
would bring near the following.

£100 £30 £10

**Red Sonja, No. 1, 1976**
Girl pin-up and a she-devil with a sword. Published by
Marvel Comics.

£50 £10 £5

**Red Sonja, No. 5, 1977**
Female equivalent of Conan. Published by Marvel
Comics.

£30 £5 £2

**The Shield Comic, 1966**
A good year for even the one-off comic hero: the Shield
took on the Black Hood and won. Published by Radio
Comics.

£100 £20 £5

**Showcase Book, No. 1, 1957**
Successful series with space stories and pictures.
Published by National Periodical Publications.

£100 £30 £5

**Showcase Books, No. 6, 1963**
Challengers of the Unknown. Cover by Jack Kirby.

£50 £20 £5

**Showcase Books, Nos. 8, 13 and 14, 1959**
The Flash was introduced in No. 8, and further
adventures followed in Nos. 13 and 14, when the Flash
was given a magazine of his own. The whirlwind
adventures of the fastest man alive, able to go where no
man had gone before. John Broome became the
scriptwriter.
No. 8

£100 £30 £10
No. 13 £50 £10 £5
No. 14 £50 £10 £5

**Showcase Book, No. 24, 1960**
Thanks to editor Julius Schwartz, another great hero of
DC Comics, the Green Lantern was given a comeback
until he received a magazine of his own. Cover by Gil
Kane.

£100 £30 £5

**Showcase Presents, Nos. 41–2, 1962**
This hero of the Planeteers had been a backup for several
earlier comic stars, but No. 41 with cover art by Lee Elias
and No. 42 gave Tommy his own top-stardom comic spot.
Published by National Periodical Publications.

£200 £40 £10

| COMICS AND MAGAZINES | MC | GC | FC |
|---|---|---|---|
| **Showcase Presents Doctor Fate and the Hourman, 1965** <br> Delightful to collect. Published by National Periodical Publications. | £200 | £40 | £10 |
| **Showcase Presents Inferior Five, No. 62, 1966** <br> Inspired by the Batman TV series, it was a well-received creation. Published by National Periodical Publications. | £200 | £40 | £10 |
| **Silver Streak Comic, No. 1, 1941** <br> With the character created by Jack Binder, Captain Battle became a legend in America. | £100 | £30 | £5 |
| **Smash Comic, No. 1, 1941** <br> This popular comic introduced some fine characters which included the Ray Man drawn by the talented Lou Fine. Published by Comics Magazine of America. | £200 | £40 | £5 |
| **The Spectre, No. 1, 1967** <br> The Spectre had been a mainstay of the DC Comics superhero line-up in the 1940s, and here he was revamped and did well. Cover by Neal Adams. | £200 | £30 | £10 |
| **Spider, No. 1, 1933** <br> A first-ever idea. No one at the time realised just how great the Spiderman to follow in later years was to become. Rare to find mint. | £10000 | £250 | £50 |
| **Spider, Master of Men, No. 1, February 1938** <br> This extra-long comic issue contained the story of 'The City of Lost Men'. Price US19c. Rare to find mint and a very good investment. | £5000 | £250 | £50 |
| **Spider Woman, No. 1, 1979** <br> Spider Woman cut a pretty figure as she fought and won millions of fans all over the world. Artists were John Romita and Tom Palmer. Published by Marvel Comics. | £50 | £10 | £5 |
| **Spiderman, Nos. 96–8, 1971** <br> Three issues with the story of a drugs plot. Because drug stories were prohibited, the editor was refused permission to publish. However, Marvel Comics went ahead and published them. Many copies were destroyed, yet some survived for the collector. | £5000 | — | — |
| **Spyman, No. 1, September 1966** <br> Colourful character worth having in a collection. Published by Harvey Comics. Good investment. | £100 | £30 | £10 |
| **Startling Comics, No. 1, 1941** <br> With Captain Future, a top space hero. | £200 | £40 | £10 |

**Steel Sterling, Man of Steel, No. 1, 1967**
Published by Radio Comics. Not very successful, but the
first issue is still rare.

| | MC | GC | FC |
|---|---|---|---|
| Steel Sterling, Man of Steel, No. 1, 1967 | £500 | £30 | £10 |

**Strange Adventures, No. 1, 1950**
Was not offensive and avoided undue violence, yet held
the reader in suspense with its tales of outer space.
Published by National Periodical Publications. £500 £50 £10

**Strange Adventures, No. 3, 1953**
With 'Stranger from Mars', by Gardener Fox and other
tales. £200 £30 £10

**Strange Adventures, No. 9, 1959**
With 'The Origin of Captain Comet'. £100 £10 £5

**Stuntman Comic**
Cover by Simon and Kirby. Published by Harvey
Features Syndicate.
No. 1, 1946 £250 £50 £10
No. 2, 1946 £150 £30 £5

**Sun Girl Comic, No. 1, 1948**
Beauty and space with partly clad females were brought
out to give comics a boost. Sun Girl fought the menace of
the monsters. Published by Prime Publications. £200 £50 £10

**Super TV Heroes, No. 1, by Hanna Barvers, 1968**
Gold Key issues and this number one is a rare find. £100 £30 £10

**Superboy, No. 1, March/April 1949**
Superboy first appeared in *More Fun Comic* in 1944, but
this was his own comic some years later. Published by DC
Comics. £200 £50 £10

**Superman Comic, No. 21, 1960/63**
The issue which contained the battle between Super-Lois
and Super-Lana, with Superman trying to stop them. £200 £30 £5

**Superman Pal, No. 70, July 1963**
This issue contained the following words of Jimmy Olsen:
'Nobody, but nobody, will guess the Secret of the Silver
Kryptonite'. Published by National Periodical
Publications. £100 £30 £5

**Supersnipe Comic, No. 1, 1944**
Ed Gruskin came up with a winner in Muscleman Kappy
McFad with designs by C.M. Payne leading a fine team.
Published by Street & Smith Publications. £100 £30 £5

**Tales of the Mysterious Traveller, 1956**
The success just kept on going in this boom year.
Published by Charlton.    £100    £30    £5

**Tales to Astonish, No. 1, 1961**
More heroes to add to the world of space and mystery.
Published by Vista Publications.    £100    £30    £5

**Tales to Astonish, No. 23, 1963**
With Moomba. Cover art by Jack Kirby and Dick Ayers.    £50    £10    £5

**Teenage Mutant Ninja Turtles, No. 1, 1984**
This is one of the most incredible success stories one could
ever read. Quite honestly the title of this series should be
'Teenage Mutant Ninja Millionaires'. Born in a sewer in
1984, the Turtles went on to become one of the most
widely merchandised comic book ventures and
properties in the world. There were four of them:
Donatello, Leonardo, Michelangelo and Raphaelo.
Created by writer Kevin Eastmand and artist Peter Laird.
Any person with a copy signed by the creators has a
terrific investment indeed. Normal value.    £200    £50    £10

**Thrilling Comic, No. 1, 1947**
With Nazis being attacked from space, a comic which
won many fans. Published by Standard Magazines.    £500    £50    £10

**Thunder Agents, No. 1, 1966**
Published by Tower Comics, who produced some of the
best comics that ever appeared on the market.    £100    £30    £10

**Top-Notch-Laugh Comic, No. 1, 1942**
With many space characters making readers laugh and
have fun. Created by Bob Montana. Published by MLJ
Magazines.    £100    £30    £5

**Vigilante, No. 1, 1983**
Violent man of action. Cover for his debut by Keith
Pollard. Published by DC Comics.    £200    £30    £10

**Weird Science Book, No. 1, 1952**
William Gaines was one of the top science fiction people.
Wally Wood did some great covers for the magazines.
Published by Fables.    £500    £50    £10

**Weird Worlds, No. 1, 1973**
Special by Edgar Rice Burroughs, the well-known author
of the Tarzan series, it featured the adventures of David
Innes and John Carter of the planet Mars.    £50    £20    £5

| | MC | GC | FC |
|---|---|---|---|
| **Whirlwind Comic, No. 1, 1940**<br>With the great Cyclone, successful in all ways to a winning comic. Published by Bilbara. | £100 | £40 | £10 |
| **Whiz Comic, No. 1, 1942**<br>With Ibis the Miracle Man of Invincible Strength. Price US10c. Published by Fawcett Publications. | £200 | £50 | £10 |
| **Wings Comic, No. 1, 1948**<br>Artist and creator Bob Lubbers gave the world Captain Wings and the Red Rocket, while at the same time showing plenty of leggy high-flyers. Published by Fiction House Magazines. | £200 | £50 | £10 |
| **Wonder Comic, No. 1, May 1939**<br>This issue saw the first and last appearance of a new hero, Wonder Man. Because of his very close resemblance to Superman – Fred's only difference was his all-red costume – legal action was taken. Good investment. | £1000 | £100 | £20 |
| **Wonder Woman, No. 1, Summer 1942**<br>Without any doubt Wonder Woman became the most successful female character in the history of all comics. Published by DC Comics. | £200 | £50 | £10 |
| **Wow Comic, No. 7, 1942**<br>Very soon after Batman and Robin became famous, other Boy Wonders arrived in different comics. This comic featured Scarlet's Pinky and the Phantom Eagle. Published by Fawcett Publications. | £50 | £20 | £5 |
| **The X-Men, No. 1, 1965**<br>This series introduced the idea of mutant heroes into comic books. This monthly magazine brought together in time the Iron Man, Thor, the Hulk, Ant Man, the Wasp and Captain America. Price US12c. Published by Canam Publishing Co. | £1000 | £200 | £30 |
| **The All-New X-Men, No. 1, 1978**<br>Featuring Wolverine the Savage, this re-shuffle was more successful than *The X-Men*. Published by Marvel Comics. | £30 | £10 | £3 |
| **Young Allies Comic**<br>Colourful, adventuresome and futuristic. Cover by Simon and Kirby, who had a cult following. Published by USA Publications. | | | |
| No. 1, 1941 | £200 | £40 | £5 |
| No. 8, 1943 | £100 | £20 | £5 |
| **Zip Comic, No. 1, 1940**<br>With the Man of Steel, successful in every way. By Charles Biro. Published by MLJ Magazines. | £100 | £30 | £5 |

# OTHER
# PUBLICATIONS

| | MC | GC | FC |
|---|---|---|---|
| **Action Toy Book, by James Razzi** | | | |
| Contains eight cut-out toys in brightly coloured laminated card. Put them together and watch them move; no scissors or paste needed. Make your Ray Gun, Klingon Cruiser, Tricorder, Phaser-Universal Translator Communicator, USS Enterprise and Vulcan Ears. Price US$2.95. Published 1976 by Random House. | £50 | £20 | £10 |
| **The Adventures of Robot Ha Ha, by Eduardo Paolozzi and others** | | | |
| Published in the 1950s. | £30 | £20 | £10 |
| **The Adventures of Tin-Tin the Shooting Star** | | | |
| Published 1947 by Editions Casterman, Paris and Tournai. Translations have sold by the tens of thousands, but the original is what brings the top prices. | £100 | £30 | £10 |
| **Amazing Stories** | | | |
| Includes 'Can Earth Repel an Alien Invader' and 'The Iron Men of Venus' by Don Wilcox. Published 1952. | £30 | £20 | £10 |
| **Arnold of Germany Catalogue** | | | |
| This firm made some excellent items and their published adverts are also worth collecting. | £30 | £20 | £10 |
| **Asakusa of Japan Catalogue** | | | |
| Early editions are worth more. | £20 | £10 | £5 |
| **Bash, by Eduardo Paolozzi** | | | |
| Published 1971. Good and rare find. | £40 | £20 | £5 |
| **Biliken Toy Company of Japan Catalogue** | | | |
| This firm made some good space toys and the editions of their fine catalogues are worth collecting. | £30 | £10 | £5 |
| **Biller of Germany Catalogue** | | | |
| Rare models are easier to find through such catalogues, regardless of date of issue, although catalogues from before the Second World War are rare and worth more. | £50 | £20 | £10 |

**Britain Catalogue**
With the many wonderful models made by this firm the catalogues are good investments, especially the 1982 edition with space models on pages 2 and 3. Prices start as I advertise, but they can go up and down according to quality and how many are available. £30 £15 £5

**Buck Rogers Better Little Book**
Buck Rogers in the war with little Venus, a planet that many people have written about. Published 1938 by Whitman Publishing. £100 £40 £5

**Captain Scarlet Official Set**
Play pack containing colouring book, stories, pad and crayons. Price £1.99. Published 1993 by Copyright Promotions. Must be in celo-wrapper. £15 £5 £2

**Chad Valley Toys**
This fine, traditional firm has good credit to its name and was the pride of many collectors. Catalogues include space toys. £50 £20 £10

**The Cloud Atomic Laboratory, by Eduardo Paolozzi and others**
Published 1956. £30 £20 £10

**Cloud Atomic Laboratory**
Special edition published 1958. £50 £10 £5

**The Conditional Probability Machine**
Images of space creatures. Published 1956. £30 £20 £10

**Conn of America Catalogue**
Rare. £50 £30 £10

**Corgi Catalogue**
Toy firm still going strong, Many of their first catalogues are rare. Not all the catalogues have information about space toys. £50 £20 £10

**Crusader Rabbit**
Children's nursery type of book, full of very interesting characters and feats and an outer planet type with a dragon called Friendly. Published 1957 by Dell Books. £100 £30 £10

**Denys Fisher Catalogue**
This company had a wonderful factory at Wetherby, near Leeds. One of the best English firms, especially for space toys, it was a great pity when they closed their doors for the last time in 1980. Their rare catalogues of 1978 bring high prices. £100 £50 £25

**Dinky Catalogue**
Several catalogues are very rare, including the fine and
last edition of 1978 in which the details were changed by
hand as the firm fell into receivership. Prices vary.

£50 £25 £15

**Dunbee-Combex Model Catalogue**
This firm made many take-overs during its existence. It
had several top space toys.

£30 £10 £5

**Fan Publications**
Any of the following are worth adding to your collection
giving interesting information for early starters in the
comic world: *The EC Fan Bulletin; The EC Fan Journal,
Concept; The EC Scoop; The EC World Press;* etc.

Negotiable

**Fantastic Adventures**
Published March 1939.

£50 £20 £10

**The Female Robot's Arrival**
Published 1952. Very rare.

£100 £50 £20

**Future Science Fiction**
Published November 1953.

£30 £20 £10

**Galaxy of Science Fiction, Vol. 3, No. 5**
'The Defenders' by Philip K. Dick.

£50 £25 £10

**Giant in the House**
Pop-up book. Published by Random House.

£40 £20 £10

**The Gladiator, by Philip Wylie**
This has definitely been recognised as the book which
gave this writer the ideas to bring one of the greatest
space figures of all time to life.

£1000 £200 £50

**Graphic Story Magazine, No. 13**
This issue contains 'Adam Link's Vengence', 'War
Machine', 'Mal-ig' and an interview with John Severin.
US$1.25.

£50 £25 £10

**Herge Casterman**
Published 1952.

£30 £10 £5

**The Japanese War God, by Eduardo Paolozzi**
Published 1958.

£30 £20 £10

**Judgement Day**
About Joe Orlando the spaceman. Published 1953.

£50 £20 £10

**Linemar Catalogue**
Another catalogue to collect regardless of year of issue.

£20 £10 £5

| --- | --- | --- | --- |

**Lone Star Catalogue**
From their base in Hatfield, Hertfordshire, this firm made
a fine selection of models, including space toys.

|  | £30 | £20 | £10 |

**Love and Rockets, by Gilbert and Jaime Hernandez**
Published in 1982 amid a debate about the growing sex
and violence in comic books. In a survey of more than 100
titles only three or four were passed as being legally fit
for publication.

|  | £20 | £5 | £2 |

**The Mandrake**
Book on robots and stories of bravery. Published in 1936.

|  | £100 | £40 | £20 |

**Matchbox Catalogue**
Many fine editions have been issued since the early days
in the 1960s. Prices vary. Features space toys.

|  | £50 | £20 | £10 |

**The Meccano Magazine**
One of the very rare issues which hit the model shops in
the years immediately after the Second World War.
Features space toys.

|  | £50 | £30 | £10 |

**Metal Huriant (First Series)**
Published in a series and price is according to what you
have.

|  | £50 | £25 | £10 |

**Metal Huriant (Second Series)**
With special 100 pages.

|  | £50 | £25 | £10 |

**Mettoy Catalogue**
This firm was successful and some catalogues are very
rare, even the trade issues. Features space toys.

|  | £50 | £20 | £10 |

**Mickey and the Robots**
Published 1952 by Hachette.

|  | £100 | £30 | £10 |

**The Monster and the Ape**
Also a movie. Published 1945.

|  | £50 | £30 | £10 |

**The Phantom Creeps**
Published 1952. Rare.

|  | £100 | £50 | £20 |

**Potrzebie, by Bob Stewart**
Published 1954.

|  | £50 | £10 | £5 |

**Prisoner of Vega**
Good story book. Published by Random House.

|  | £30 | £15 | £5 |

| | MC | GC | FC |
|---|---|---|---|
| **The Puzzle Manual, by James Razzi**<br>Includes mazes, puzzles and trivia for all ages as well as space age. Price US$5.95. Published 1976 by Bantam Books. | £40 | £20 | £10 |
| **Robbie the Robot**<br>From *The Day the Earth Stood Still* by Gerard Perron. | £50 | £30 | £10 |
| **Robbie and the Planet Interdite**<br>Another rare book on space. Published 1956. | £100 | £50 | £20 |
| **The Robot in the Metropolis**<br>More adventures of space. Made into a film by Fritz Lang. Published 1956. | £50 | £10 | £5 |
| **Rock Valley Toys of Japan Catalogue**<br>Earlier editions will be worth more. | £20 | £10 | £5 |
| **Schuco of Germany Catalogue**<br>Well known for many years, respected and also greatly admired. Good catalogues to have. | | | |
| Prewar | £200 | £50 | £25 |
| Early postwar | £50 | £25 | £10 |
| Later | £20 | £10 | £5 |
| **Scoops, No. 1 of Amazing New Wonder Weekly**<br>With free gift of great new puzzle, 1,000 gifts to readers for solving. Front cover features a spaceman carrying a human in nightware. Price 2d. Published 1934. | £50 | £30 | £10 |
| **Space Toys by Wako of Japan**<br>A good catalogue to own, regardless of the year it was issued. | £20 | £10 | £5 |
| **Star Wars**<br>There are several Star Wars books. Prices vary according to what you have, but realistic prices can be taken from what is given here. | £50 | £20 | £10 |
| **Startling Stories**<br>Featuring 'The Shadow Men', an astonishing new novel by E.A. Van Vogt. The return of Captain Future. Published 1950. | £40 | £20 | £10 |
| **Startling Stories**<br>Contains 'A History of Robots' by Williamson. Published 1955. | £100 | £50 | £25 |
| **Steel Jim Brick Bradford**<br>Published 1939. | £100 | £30 | £10 |

**The Story of Superman, by Ted White**
First-class investment special, a 22-page 4 x 6in.
mimeographed booklet. The first title in a series is always
important, but there were four booklets in all, even
though it was a one-off publication ideal for collectors.
Published 1952.

| | | | |
|---|---|---|---|
| First title | £250 | £40 | £10 |
| Set of four | £500 | £50 | £20 |
| Signed | Negotiable | | |

**The Facts behind Superman, by Ted White**
New title for the previous *Story of Superman*. Published
1953.　　　　£50　£10　£5

**Texas Toys Expansion Module Chatter Book**
Made specifically for 5 to 7-year-olds, a plug-in expansion
module plus. Five learning-to-read rooks which add
more practice up to a complete reading book with
comprehension questions to increase the child's confidence and experience in reading, including space.　£10　£5　£2

**Time, December 1952**
This issue of 'the weekly newspaper magazine' includes
a special space feature, 'Space Pioneer: Will Man
Outgrow the Earth?'.　　£30　£10　£5

**Tom Corbett Push-Out Book**
Fine book. Published 1952 by Saalfield, Akron, Ohio,
USA.　　£100　£30　£10

**Trillions of Things**
Pop-up book. Published by Random House.　£40　£20　£10

**The Trouble with Tribbles, by David Gerrold**
The ups and downs, as well as the ins and outs, of the
making of the very popular Tribbles episode. Price
US$1.95. Published 1976 by Ballantine Books.　£50　£25　£15

**The Truth Machine**
Star story. Published by Random House.　£30　£15　£5

**Two Renderings of the Same Image**
Very rare and interesting book on space. Published 1955.　£50　£30　£10

**Wonder Books Battlestar Galactica, by Manny
Campans**
This book on space contains photographs of stars like
Lorne Greene who played Commander Adama in this
very popular TV series. An Interstellar Activity
Publication, it also had pictures and puzzles of monsters
from space. Published 1978 by Grosset & Dunlap, a

filmways company, in conjunction with Paramount
Studios Hollywood. The book was not expensive, and
there were times when you could not give it away, let
alone sell it. Further, if you have the author's signature,
you have a winner.

|  | MC | GC | FC |
| --- | --- | --- | --- |
| Unsigned | £100 | £50 | £10 |
| Signed | £1000 | £500 | £10 |

**Zeno Energy Experimental, by Eduardo Paolozzi**

| | MC | GC | FC |
| --- | --- | --- | --- |
| Published 1958. | £30 | £20 | £10 |

# PHONECARDS

| Country and description | Face value (or units) | Mint | Used |
|---|---|---|---|
| **UK: BRITISH TELECOM** | | | |
| **Action Man** | 5 units | £10 | £2 |
| **Captain Scarlet and the Mysterons** set in special brochure  each | 5 units | £100 | — |
| **Dan Dare** | 5 units | £10 | £2 |
| **Doctor Who and the Daleks** set in special brochure issued in limited edition by the BBC 1995 | | £75 | £25 |
| **Earth Orbiter** | 5 units | £25 | £5 |
| **Fireball XL5** | 5 units | £10 | £2 |
| **Isle of Man Earth Station** | 100 units | £25 | £5 |
| **Joe 90** set of four  each | 5 units | £50 | £10 |
| **Judge Dredd** | 5 units | £10 | £2 |
| **NASA Apollo** | 5 units | £25 | £5 |
| **P and J Aliens** | 5 units | £10 | £2 |
| **The Prisoner** | 5 units | £10 | £2 |
| **Roy of the Rovers** | 5 units | £10 | £2 |
| **Shuttle Booster** | 5 units | £25 | £5 |
| **Space 1999** | 5 units | £10 | £2 |
| **Star Wars Trilogy** | 5 units | £10 | £2 |
| **Stingray** | 5 units | £10 | £2 |
| **Supercar** | 5 units | £10 | £2 |
| **Take You into Space** | 10 units | £20 | £5 |

| PHONECARDS | Face value | Mint | Used |
|---|---|---|---|
| **Thunderbird 1** | 5 units | £25 | £5 |
| **Thunderbird 2** | 5 units | £25 | £5 |

### UK: MERCURY

| | Face value | Mint | Used |
|---|---|---|---|
| **Goldfinger** | | | |
| Unsigned | £1 | £25 | £5 |
| Signed by Sean Connery | £1 | £750 | £200 |
| **Moonraker** | | | |
| Unsigned | £1 | £25 | £5 |
| Signed by Roger Moore | £1 | £500 | £250 |
| **Space Management - S** | 50 units | £20 | £3 |
| **Star Trek: Captain Kirk** | 50p | £75 | £10 |
| **Star Trek: Captain Picard** | 50p | £50 | £5 |
| **Star Trek: Doctor Crusher** | 50p | £35 | £10 |
| **Star Trek: Enterprise** | 50p | £50 | £10 |
| **Star Trek: Generations** | £5 | £50 | £15 |
| **Star Trek: Romulan Warbird** | 50p | £35 | £5 |
| **Star Trek: Tom Bob Allan 1, 2, 3, 4, 5, 6 & 7** each card | 50p | £25 | £5 |
| **Complete set in special folder** | | £200 | — |
| **Star Trek: Tom Bob Allan Special** | 50p | £35 | £10 |

### UK: P & J LIMITED EDITIONS (Issues for 1995/96)

| | Face value | Mint | Used |
|---|---|---|---|
| **Aliens 2, 3 & 4** each card | 5 units | £10 | £2 |
| **Captain Scarlet** set of four in folder each card | 5 units | £50 | £20 |
| **The Empire Strikes Back 1 & 2** each card | 5 units | £10 | £2 |
| **Fireball 2, 3 & 4** each card | 5 units | £10 | £2 |
| **Joe 90** set of four in folder each card | 5 units | £50 | £20 |

| PHONECARDS | Face value | Mint | Used |
|---|---|---|---|
| **The New Judge Dredd** set of 18 by the creator John Wagner and the artist Kev Walker in folder each card | 5 units | £200 | £50 |
| **Return of the Jedi 1 & 2** each card | 5 units | £10 | £2 |
| **Star Wars 2** | 5 units | £10 | £2 |
| **Stingray** set of four in folder each card | 5 units | £50 | £20 |
| **Supercar 2, 3 & 4** | 5 units | £10 | £2 |
| **UFO** | 5 units | £10 | £2 |

## AUSTRALIA

| | Face value | Mint | Used |
|---|---|---|---|
| **Apollo Custom** | A$2 | £45 | £10 |
| **Apollo Custom** | A$5 | £45 | £10 |
| **Apollo Custom** | A$10 | £55 | £15 |
| **Apollo Custom** | A$20 | £65 | £20 |
| **Apollo Custom** | A$50 | £100 | £30 |

## CAMBODIA

| | Face value | Mint | Used |
|---|---|---|---|
| **Planet Earth** | 2 units | £25 | £5 |

## FINLAND

| | Face value | Mint | Used |
|---|---|---|---|
| **Cosmic Ray** | FM10 | £25 | £5 |
| **Outer Space Fan** | FM30 | £25 | £5 |

## FRANCE

| | Face value | Mint | Used |
|---|---|---|---|
| **Starnights** pair each | 50 units | £30 | £10 |

## NEW ZEALAND

| | Face value | Mint | Used |
|---|---|---|---|
| **Satellite 1990** | NZ$2 | £25 | £5 |
| **Satellite 1990** | NZ$5 | £35 | £10 |

| PHONECARDS | Face value | Mint | Used |
|---|---|---|---|
| **Satellite 1990** | NZ$10 | £50 | £10 |
| **Satellite 1990** | NZ$20 | £65 | £15 |
| **Compete set of above four** | | £100 | — |

## SWITZERLAND

| | | | |
|---|---|---|---|
| **Arthur C. Clarke** 50th anniversary | 5 units | £25 | £5 |
| **Man in Space** | 10 units | £25 | £5 |
| **Marvel Cards** global set   each | 10 units | £50 | £10 |

## USA

**25th Anniversary of NASA's Lunar Landing**
Super quality set of four packed in its own
presentation case. Includes the world's first
silver telephone card and four Apollo II
embroidered patches as worn by the astronauts.
Limited issue 1969 by P.M. Powell Associates,
America's largest telecard dealer. Price $169.69.                    £25          £75

**July 1995**

Action Comic, No. 1
(see page 153)

now valued at
£50,000